Springer Tracts in Modern Physics
Volume 139

W0055305

Springer-Verlag Berlin Heidelberg GmbH

Springer Tracts in Modern Physics

Covering reviews with emphasis on the fields of Elementary Particle Physics, Solid-State Physics, Complex Systems, and Fundamental Astrophysics

Manuscripts for publication should be addressed to the editor mainly responsible for the field concerned:

Gerhard Höhler
Institut für Theoretische Teilchenphysik
Universität Karlsruhe
Postfach 6980
D-76128 Karlsruhe
Germany
Fax: +49 (7 21) 37 07 26
Phone: +49 (7 21) 6 08 33 75
Email: gerhard.hoehler@physik.uni-karlsruhe.de

Joachim Trümper
Max-Planck-Institut
für Extraterrestrische Physik
Postfach 1603
D-85740 Garching
Germany
Fax: +49 (89) 32 99 35 69
Phone: +49 (89) 32 99 35 59
Email: jtrumper@mpe-garching.mpg.de

Johann Kühn
Institut für Theoretische Teilchenphysik
Universität Karlsruhe
Postfach 6980
D-76128 Karlsruhe
Germany
Fax: +49 (7 21) 37 07 26
Phone: +49 (7 21) 6 08 33 72
Email: johann.kuehn@physik.uni-karlsruhe.de

Peter Wölfle
Institut für Theorie
der Kondensierten Materie
Universität Karlsruhe
Postfach 69 80
D-76128 Karlsruhe
Germany
Fax: +49 (7 21) 69 81 50
Phone: +49 (7 21) 6 08 35 90/33 67
Email: woelfle@tkm.physik.uni-karlsruhe.de

Thomas Müller
IEKP
Fakultät für Physik
Universität Karlsruhe
Postfach 6980
D-76128 Karlsruhe
Germany
Fax:+49 (7 21) 6 07 26 21
Phone: +49 (7 21) 6 08 35 24
Email: mullerth@vxcern.cern.ch

Roberto Peccei
Department of Physics
University of California, Los Angeles
405 Hilgard Avenue
Los Angeles, California 90024-1547
USA
Fax: +1 310 825 9368
Phone: +1 310 825 1042
Email: robertop@college.ucla.edu

Frank Steiner
Abteilung für Theoretische Physik
Universität Ulm
Albert-Einstein-Allee 11
D-89069 Ulm
Germany
Fax: +49 (7 31) 5 02 29 24
Phone: +49 (7 31) 5 02 29 10
Email: steiner@physik.uni-ulm.de

James Hamilton

Aharonov–Bohm and other Cyclic Phenomena

With 34 Figures

Springer

Prof. James Hamilton

4 Almoner's Avenue
Cambridge CBI 4 PA
England

Library of Congress Cataloging-in-Publication Data

Hamilton, J. (James), 1918 – Aharonov-Bohm and other cyclic phenomena / James Hamilton. p. cm. - (Springer tracts in modern physics, ISSN 0081-3869; v. 139) Includes bibliographical references and index.
1. Quantum theory. 2. Cycles. I. Title. II. Series: Springer tracts in modern physics; 139. QC1.S797 vol. 139   [QC174.12]   539 s–dc21   [530.12]   97-22633   CIP

Physics and Astronomy Classification Scheme (PACS): 03.65.B, 72.15.R

ISBN 978-3-662-14805-1   ISBN 978-3-540-68419-0 (eBook)

DOI 10.1007/978-3-540-68419-0

© Springer-Verlag Berlin Heidelberg 1997

Originally published by Springer-Verlag Berlin Heidelberg New York in 1997.

Softcover reprint of the hardcover 1st edition 1997

Typesetting: A. Leinz, Karlsruhe
Cover design: *design & production* GmbH, Heidelberg
SPIN: 10546202   56/3144-5 4 3 2 1 0 – Printed on acid-free paper

Preface

The Aharonov–Bohm effect may seem to be strange. The interference pattern of an electron wave, in a region with no electromagnetic field present, depends on the static value of a magnetic flux elsewhere. Moreover the effect cannot be detected for a single electron; it takes a number of electrons, or repeated experiments with one electron, to demonstrate it. The quantum theory limit on the accuracy of measurement of flux is relevant here.

The Aharonov–Bohm effect is a simple consequence of applying wave mechanics to electromagnetic theory, and there are numerous practical examples in which it, or a generalization of the effect, appears. Various works criticise in principle the original Aharonov–Bohm effect, or else they criticise the experiments by which it was verified.

It seemed to be worthwhile to give a coherent account of these fairly simple, but basic, developments in atomic theory. It was suggested to me that it was sensible to add a survey of other effects – not necessarily in atomic physics – which also are related to motion in cycles, in cases where the dynamics is not symmetric under the reversal of time. Such are Foucault's pendulum, Berry's phase, the Sagnac effect, etc. These are examples of anholonomic phenomena.

This volume is intended primarily for graduate students and research workers, but it may also be of value to those who are interested in the general laws of physics.

Cambridge, February 1997 *J. Hamilton*

Contents

1. Introduction

1.1 General Remarks

The purpose of this monograph is to discuss the nature of the Aharonov–Bohm effect and to consider also some other effects that are associated with cyclic motion. The common feature is that the phenomena are anholonomic. This means that the dynamical situation of a system depends not only on the general coordinates at the current position but also depends on the route by which the system reached the current position. Thus Foucault's pendulum at noon today will, in general, move in a different plane than it did at noon yesterday.

The systems we consider, including the Aharonov–Bohm effect, have the property that they violate Wigner's type of time reversal invariance. The part of the solution that arises from this violation is usually of particular interest.

The Aharonov–Bohm effect can be generalised by considering the motion of an electron in a uniform induction field B. It can be further generalized by looking at motion on the circle $\rho = \rho_T$ in the cylindrical symmetric induction field of the form

$$B_z = B(\rho/\rho_T)^{n-1} \quad ,$$

where B and n $(n \geq 0)$ are constants, and ρ is the distance from the axis of symmetry Oz. The anholonomic effect can easily be seen in this way. If is not only the induction on the electron's path that matters, as can be seen in the design of accelerators.

In the first part (Chaps. 1–7) there is an account of some of the criticisms of the Aharonov–Bohm effect, and of the arguments about its dynamical origin. The measurements of the effect are studied, as well as its role in superconducting phenomena. There is a critical account of the basic quantum limitations on the accuracy of any flux measurement.

The second part (Chap. 8) discusses adiabatic theorems, various anholonomic phenomena, Hannay's angle and Berry's phase. There are applications to Larmor precession, accelerator physics, and to the relation between the Aharonov–Bohm phase, Berry's phase, and a new quantity called the γ-phase.

Various useful and relevant mathematical details are given in Appendices A to H.

1.2 The Aharonov–Bohm Effect

W. Franz in 1939 wrote on the effect of magnetic flux on electron interference [1.1] and in 1949 W. Ehrenberg and R. E. Siday [1.2] wrote a pioneering paper on the motion of an electron in a magnetic field. In 1959 Y. Aharonov and D. Bohm [1.3] wrote about the phase factor that became known as the Aharonov–Bohm effect (we shall use "AB effect" for convenience). In 1960 R. G. Chambers [1.4] reported an experimental verification of the effect, and together with M. H. L. Pryce [1.4], discussed the theoretical interpretation of the tapered whisker fringes. Good surveys of various topics relevant to the AB effect are to be found in [1.1]–[1.7].

There has subsequently been a fair amount of confusion about the nature of the AB effect. Much of the confusion can be traced to a misunderstanding of certain aspects of quantum theory. Two points are worth making here:

(a) In the case of the AB effect a non-local boundary condition relates the change in the phase of an electron wave function (on going once around a solenoid) to the amount of flux Φ in the solenoid. This applies regardless of whether there is a considerable probability that the electron is close to the solenoid, or only a very small probability. That is quite unlike a classical interaction, and so the consequences may seem to be strange. There is further discussion in Sect. 5.3.

(b) It is shown in Sect. 6.5 that the AB effect cannot be detected for one electron, or even for a few electrons. It requires many electrons, or other charge carriers, to demonstrate the AB effect clearly. This is in strong contrast to, for example, the Lorentz force on an electron; a single electron's passage can show the latter clearly.

Thus points (a) and (b) independently make it clear that we are dealing with a purely quantum phenomenon that may well be of unfamiliar nature.

In Sect. 2.7 the important physical relation between the AB effect and Faraday's electromagnetic induction is discussed. In Sect. 2.8 the relation of the AB effect to constant acceleration in a quantum system, and to the propagation of electrons in a (periodic) metallic conductor is examined.

An interesting feature of the AB effect is that it can be generalized beyond the solenoid case. If an electron moves perpendicular to the field in uniform induction, it will describe a circle. It is shown in Chap. 3 that there is again an extra phase factor, and in this case it depends on the flux enclosed by the orbit. In Sect. 3.6 the result is extended to a hollow axial symmetric induction, in which the enclosed flux differs from that in the uniform case. In all these cases a basic feature is the use of a single-valued wave function.

In Chap. 4 the origin of the AB formalism is traced to classical physics. The angular momentum around an axis has the same dimensions as the action variable for motion around the axis. The action variable is a constant of motion, and the difference between it and the angular momentum is a constant times the AB phase angle. The examples given above are trivial in the sense that they are, or can be, circular motion; however, in the classical treatment it is easy to see the effect of the Lorentz force which is in general a non-central force.

In Chap. 5 the standard biprism experiment is examined, and a distinction is made between detecting Faraday's electromagnetic induction and detecting the AB effect. Pryce's explanation of the tilted interference pattern arising from the tapered magnetic whisker is examined in Sect. 5.6. In Sect. 5.8, 9 there are discussions of the extent to which the AB effect can be interpreted as the result of fringe fields, or the extent to which it can be built up by (i) altering the flux in the solenoid, (ii) using tapered solenoids, and (iii) using truncated solenoids. Theoretical arguments on the importance of potentials versus fields are also discussed.

Section 5.10 contains an account of the interesting phenomenon known as Aharonov–Bohm scattering. This is also discussed in Appendix G. Experiments with BCS superconducters that detect, or are relevant to, the AB effect are discussed in Sects. 5.11, 12.

In Chap. 6 the problem of the quantum theory limit on the accuracy of measuring magnetic flux is discussed, together with an elementary analysis of how repeated measurements can, in principle, give much improved accuracy without violating quantum theory.

Sections 6.6–9 discuss superconducting devices and the relevant number-phase uncertainly rule.

Chapter 7 discusses phenomen similar to the usual (magnetic) AB effect, such as the electric AB effect, the Aharonov–Casher effect and the scalar AB effect.

Relevant to these chapters are Appendices AI and AII dealing with the quantum solution to motion in particular axial symmetric induction fields. In Appendix H there is an account of the pure wave mechanics solution to motion in a uniform induction field. Appendix C contains an introduction to using a toroidal coordinate system.

2. Motion Around a Solenoid

Here we discuss the solution of the wave equation for a charged particle moving around a solenoid. We point out that the Aharonov–Bohm phase factor relates two significant wave functions, and present a transformation that can be used to prove that only quantum size effects can be detected outside the solenoid.

2.1 Basic Ideas

The MKS system of electromagnetic units will be used throughout. A basic entity is the mechanical force $\boldsymbol{F}$ which acts on a charge q that is moving with velocity $\boldsymbol{v}$ in an induction field $\boldsymbol{B}$. We have

$$\boldsymbol{F} = q\boldsymbol{v} \times \boldsymbol{B} \tag{2.1}$$

(in a right-handed coordinate system). This force $\boldsymbol{F}$ points inwards on an electron circling in the positive sense about $\boldsymbol{B}$, so for an electron

$$q = -e \ , \qquad \text{with} \quad e = +1.6 \times 10^{-19}\text{C} \ .$$

Schrödinger's equation for the motion of a particle [consistent with (2.1)] follows on replacing the momentum operator $\boldsymbol{p}$ by

$$\boldsymbol{\Pi} = \boldsymbol{p} - q\boldsymbol{A} \ , \tag{2.2}$$

where $\boldsymbol{B} = \text{curl}\,\boldsymbol{A}$. The commutator of the components of $\boldsymbol{\Pi}$ is

$$[\Pi_j, \Pi_k] = iq\hbar B_l \ , \qquad (j, k, l : \text{ cyclic } 1, 2, 3) \ . \tag{2.2a}$$

An important case concerns a region of space R over which $\boldsymbol{B}$ vanishes, but the vector potential $\boldsymbol{A}$ is not zero. An example is the region $\rho > \rho_{\rm C}$ outside the infinite circular cylinder solenoid in Fig. 2.1. The induction $\boldsymbol{B}$ is uniform in $\rho < \rho_{\rm C}$ and points along the axis Oz. In terms of the cylindrical coordinates (z, ρ, θ) the vector potential for $\rho > \rho_{\rm C}$ can be written

$$\boldsymbol{A} = (0, 0, A_\theta) \ ; \qquad A_\theta = \rho_{\rm C}^2 B/2\rho = \Phi/2\pi\rho \ . \tag{2.3}$$

The quantity $\Phi = B\pi\rho_{\rm C}^2$ is the flux in the solenoid.

In space R the components of $\boldsymbol{\Pi}$ commute, so $(\boldsymbol{\Pi}, \boldsymbol{x})$ could be used as conjugate dynamical variables, just as $(\boldsymbol{p}, \boldsymbol{x})$ is used when no magnetism is

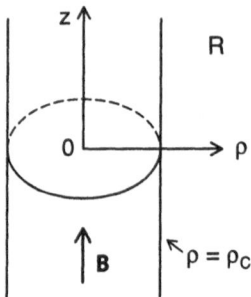

Fig. 2.1. The inifinite uniform solenoid

present. This suggests that the motion of a charged particle in R is given by $|\psi_0(\boldsymbol{x})|$ where (with suitable boundary conditions) $\psi_0(\boldsymbol{x})$ is the wave function for the case that $\boldsymbol{B} = 0$ in the solenoid. However the phase of the wave function $\psi(\boldsymbol{x})$ in R for the case in Fig. 2.1, is not the same as it would be when $\boldsymbol{B} = 0$ in the solenoid. This is the Aharonov–Bohm effect.

There is some hint of a paradox here. Various authors find it strange that in R, where there is no electric or magnetic field, the phase of $\psi(\boldsymbol{x})$ can depend on the value of $\boldsymbol{B}$ in the solenoid. But this paradox vanishes on remembering that we cannot change the induction in the solenoid from $\boldsymbol{B}_1$, to $\boldsymbol{B}_2$ without causing Faraday's electromagnetic induction. That gives an electric field encircling the solenoid, which in turn can alter the phase of the wave function $\psi(\boldsymbol{x})$ at any fixed position in R. This is indeed "action at a distance", but that action at a distance is already inherent in Faraday's induction.

There is no real difficulty in understanding how the flux in the solenoid can influence the phase of the wave function $\psi(\boldsymbol{x})$ in R. The integral of the vector potential around the solenoid,

$$\oint \boldsymbol{A} \cdot \mathrm{d}\boldsymbol{s} \ , \tag{2.3a}$$

determines the boundary condition of $\psi(\boldsymbol{x})$ on the solenoid. [The way this occurs can be seen in Sect. 2.10 below, especially in (2.35a).] It should also be remembered that the quantity in (2.3a) is gauge invariant; it is just the flux $\varPhi$ in the solenoid.

2.2 The Confining Potential

In accord with Eqs. (2.1, 2) Schrödinger's equation for an electron of mass m moving in a magnetic field is given by the Hamiltonian

$$H = -\frac{\hbar^2}{2m} \{\boldsymbol{\nabla} + \mathrm{i}(e/\hbar)\boldsymbol{A}\}^2 + V(\boldsymbol{x}) \ . \tag{2.4}$$

In general $V(\boldsymbol{x})$ is the potential energy of the electron, and it will represent conservative forces. Here it is convenient to let $V(\boldsymbol{x})$ describe a confining

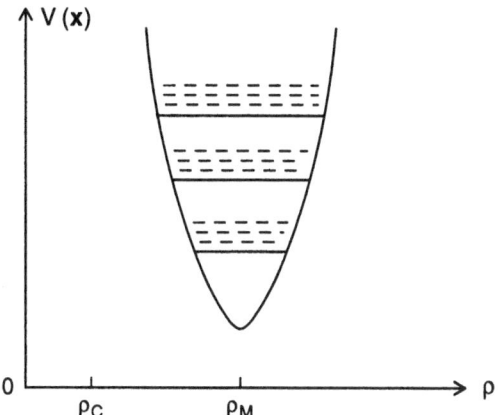

Fig. 2.2. The confining potential $V(x)$. Solid and broken lines show vibrational and rotational levels

potential. Its job is to restrain the particle to move in a chosen orbit within the region R. Figure 2.2 shows the suitable form of $V(x)$ for the case of the circular cylindrical solenoid of Fig. 2.1. The confining potential here must be a function of ρ only, and the wave function, or wave packet, will have the usual wave-mechanical spread about ρ_m. This spread can be macroscopically small, and the wave function can only penetrate very little (if at all) into the region $\rho < \rho_C$.

There may be some objection to using an infinite cylinder, as in Fig. 2.1, on the grounds that it is not a practical device. Instead, the solenoid could be wound on a torus, and the same arguments (which we shall give) would still be valid. Appendix C contains some of the mathematical tools required to treat the toroidal solenoid.

2.3 The Two Wave Functions

First we assume that the induction does not vary with time, so that $\dot{\boldsymbol{A}}(x) \equiv 0$. The actual wave function $\psi_A(\boldsymbol{x})$ obeys

$$i\hbar\frac{\partial \psi_A}{\partial t} = H\psi_A \quad . \tag{2.5}$$

For the case that $\boldsymbol{B} = 0$ in the solenoid we have

$$i\hbar\frac{\partial \psi_0}{\partial t} = H_0\psi_0 \quad , \tag{2.5a}$$

where $\psi_0(\boldsymbol{x})$ is the wave function for $\boldsymbol{B} = 0$, and

$$H_0 = -\frac{\hbar^2}{2m}\boldsymbol{\nabla}^2 + V(\boldsymbol{x}) \quad . \tag{2.4a}$$

It is easy to show that corresponding to any solution $\psi_A(\boldsymbol{x})$ of (2.5) in the region R (where $\boldsymbol{B} = 0$), there is *a solution* $\psi_0(\boldsymbol{x})$ of (2.5a), such that

$$\psi_A(\boldsymbol{x}) = U_A(\boldsymbol{x})\psi_0(\boldsymbol{x}) \quad , \tag{2.6}$$

where

$$U_A(\boldsymbol{x}) = \exp\left[-\mathrm{i}(e/\hbar)\int_{\boldsymbol{x}_0}^{\boldsymbol{x}} \boldsymbol{A}(\boldsymbol{x}')\,\mathrm{d}\boldsymbol{x}'\right] \quad . \tag{2.7}$$

In the integral over $\boldsymbol{x}'$ in (2.7) the end points $\boldsymbol{x}_0$, $\boldsymbol{x}$ and the contour along which we integrate, are all in the region R. The end point $\boldsymbol{x}_0$ is otherwise arbitrary. The function $U_A(\boldsymbol{x})$ can only exist in a region R where $\mathrm{curl}\,\boldsymbol{A}$ is zero.

Equation (2.6) is the Aharonov–Bohm relation [2.1] (see also [2.2]). Notice however that a solution of (2.5), or (2.5a), is not necessarily a wave function. In general wave functions have also to obey certain physical boundary conditions. It can happen that $\psi_A(\boldsymbol{x})$ is a wave function while the corresponding $\psi_0(\boldsymbol{x})$ is not, or vice versa.

If $\boldsymbol{x}$ starts from $\boldsymbol{x}_0$ and moves exactly once around the solenoid, in the positive direction (and in R), the phase factor of (2.7) becomes

$$\exp(-2\pi\mathrm{i}\Phi/\phi_e) \quad , \tag{2.7a}$$

where Φ is the flux in the solenoid and

$$\phi_e = h/e = 4.13 \times 10^{-15} \text{ Weber}$$
$$= 4.13 \times 10^{-7} \text{ Gauss} \cdot \text{cm}^2 \tag{2.8}$$

is the quantum unit of flux associated with the charge $(-e)$. If $\boldsymbol{x}$ moves around from $\boldsymbol{x}_0$ through azimuthal angle θ, by (2.7) the AB phase factor U_A becomes

$$U(\theta) = \exp\left(-\mathrm{i}\theta\Phi/\phi_e\right) \quad . \tag{2.9}$$

As is pointed out in Sect. 2.9 below, ψ_A and ψ_0 cannot both the single valued unless Φ/ϕ_e is an integer.

2.4 Use of Expectation Values of the Hamiltonian

The simple arguments above show that care is required in interpreting the AB relation in (2.6). In order to avoid ambiguities, the Hamiltonian of (2.4) will now be used with a single valued wave function $\psi(x)$.

Let this wave function be small outside the torus in Fig. 2.3 which encircles the cylindrical solenoid. The component of the wave function relating to the Oz behaviour can be the factor $\Xi(z)$ describing a narrow wave packet in Oz, with zero mean momentum along Oz. This wave packet will necessarily spread with time, but that can be ignored here. (Alternative a suitable confining potential $V(z)$ could be used.) The behaviour in ρ will be determined by the confining potential $V(\rho)$, and the $z - \rho$ part of the wave function will be labelled χ.

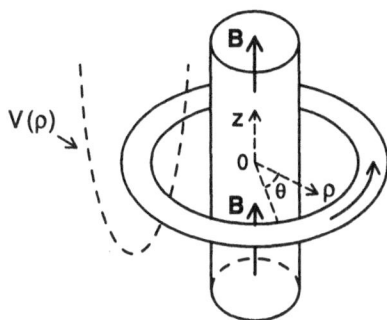

Fig. 2.3. A wave packet with some extension in the direction Oz is also confined by the potential $V(\rho)$

With the vector potential of (2.3), in $\rho > \rho_C$ Eq. (2.5) can be written

$$i\hbar\frac{\partial\psi}{\partial t} = \left[-\frac{\hbar^2}{2m}\left\{\frac{\partial^2}{\partial z^2} + \rho^{-1}\frac{\partial}{\partial\rho}\left(\rho\frac{\partial}{\partial\rho}\right)\right\} \right.$$
$$\left. +\frac{\hbar^2}{2m\rho^2}\left\{\frac{\partial}{i\partial\theta} + P(t)\right\}^2 + V(\rho)\right]\psi \quad . \tag{2.10}$$

Here

$$P(t) = \Phi(t)/\phi_e \quad . \tag{2.10a}$$

Now a slow change of $\Phi(t)$ with time is allowed, but it is subject to the condition

$$\tau|\dot{P}(t)| \ll 1 \quad , \tag{2.10b}$$

where τ is of the order of the time of rotation of a wave packet around the solenoid. Thus $P(t)$ would be approximately constant over a number of revolutions. For stationary state solutions, Φ must be constant.

Choose a solution of the form

$$\psi = \chi(z, \rho, M, \Phi, t)\exp(iM\theta) \quad , \tag{2.11}$$

where M is any integer, or zero. The function χ obeys

$$i\hbar\frac{\partial\chi}{\partial t} = \left[-\frac{\hbar^2}{2m}\left\{\frac{\partial^2}{\partial z^2} + \rho^{-1}\frac{\partial}{\partial\rho}\left(\rho\frac{\partial}{\partial\rho}\right)\right\} \right.$$
$$\left. +\frac{\hbar^2}{2m\rho^2}\left\{M + P(t)\right\}^2 + V(\rho)\right]\chi \quad . \tag{2.12}$$

It is now clear that in the state ψ of (2.11) the expectation value of the second term in square brackets on the right of (2.10) is

$$\frac{\hbar^2}{2m\rho^2}\left\{M + P(t)\right\}^2 \quad . \tag{2.12a}$$

This is the centrifugal potential energy, and it is in principle observable, as can be seen in Fig. 2.4. Different values of this term will give rise to somewhat different energies of the electron in a stationary state.

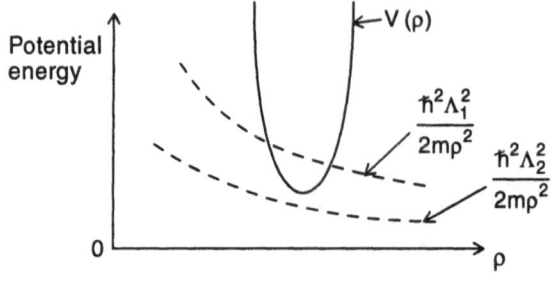

Fig. 2.4. The confining potential $V(\rho)$ and the centrifugal potential for quantum numbers $\Lambda_1 > \Lambda_2$

It follows that the operator Λ_z defined by

$$\Lambda_z = L_z + \hbar P(t) \tag{2.13}$$

is the component of the electron's angular momentum about Oz, in the case of states having the form in (2.11). Here

$$L_z \equiv \frac{\hbar \partial}{i \partial \theta} \quad . \tag{2.13a}$$

Notice that, analogous to Eq. (2.2), we could write

$$\boldsymbol{\Lambda} = \boldsymbol{x} \times \boldsymbol{\Pi} = \boldsymbol{L} - q\boldsymbol{x} \times \boldsymbol{A} \quad . \tag{2.13b}$$

Using (2.3) and $q = -e$, this again gives (2.13) in R.

2.4.1 Equipment

A particular piece of equipment, such as the circular cylindrical solenoid of Fig. 2.1, plus the appropriate confining potential of Fig. 2.2, will be called the *apparatus*. Another example would be the toroidal solenoid of Appendix C, plus its confining potential.

The value of the magnetic induction B in the solenoid (or the value of the flux Φ) will be called the *setting* of the apparatus. It can be operated at various settings; or the setting may be altered slowly with time.

2.4.2 Eigenvalues

Considering the apparatus in Fig. 2.3 it follows from (2.13) that in R, a solution ψ of the form in (2.11) has the eigenvalues Λ and $M\hbar$ for Λ_z and L_z respectively, where

$$\Lambda = M\hbar + \frac{e}{2\pi}\Phi \quad . \tag{2.14}$$

Notice that unless Φ is an integral multiple of ϕ_e, Λ does not have eigenvalues which are integral multiples of $\hbar$. However, the form in (2.11) is required in order that ψ be single valued; and that choice has clear advantages. The forms in (2.11) and (2.14) also make it easier to understand the dynamics, as we now show.

L. J. Tassie and M. Peshkin [2.3] have emphasised the importance of using the single-valued wave function [as in (2.11) above], even in a multiply connected region, for treating the AB effect with Schrödinger's equation.

2.5 The Flux Transformation

Solutions of (2.10) for different settings can be simply related. Choose any (positive or negative) integers M, M'. The flux transformation is defined as

$$M \longrightarrow M'$$
$$\Phi(t) \longrightarrow \Phi'(t) = \Phi(t) + (M - M')\phi_e \quad .$$
(2.15)

Take any solution χ of (2.12) for a range of values of $\Phi(t)$. By Eqs. (2.12) and (2.15) the *same solution* χ can be used in the two cases:

$$\psi_1 = \chi(z, \rho \, M, \Phi(t), t) \exp(iM\theta) \quad ,$$
$$\psi_2 = \chi(z, \rho \, M', \Phi'(t), t) \exp(iM'\theta) \quad .$$
(2.15a)

The wave functions ψ_1 and ψ_2 obey (2.10); also they have the same eigenvalue Λ for Λ_z, where

$$\frac{2\pi}{e}\Lambda = \Phi(t) + M\phi_e = \Phi'(t) + M'\phi_e \quad .$$

Thus each solution χ of (2.12) in R gives rise to a family of (equivalent) solutions ψ that relate to different settings $\Phi(t)$ of the apparatus. We have seen that any pair of such solutions, ψ_1 and ψ_2 cannot be distinguished by using Λ_z.

Another relevant operator is the azimuthal current density j_θ. With the sign in (2.4), the current density for electrons is given by

$$j = -\frac{\hbar e}{2mi}(\psi^* \boldsymbol{\nabla}\psi - \psi\boldsymbol{\nabla}\psi^*) - \frac{e^2}{m}\psi^* \boldsymbol{A}\psi \quad .$$

In R the expectation value of the azimuthal component is

$$\langle j_\theta \rangle_R = -|\chi|^2 e\Lambda/m\rho \quad ,$$
(2.15b)

and this is the same for ψ_1 or ψ_2. (For simplicity of presentation, in (2.15b) ρ^{-1} is used for the mean value of ρ^{-1} under the influence of the confining potential $V(\rho)$ of Fig. 2.4.)

It is shown in Sect. 5.1 below that the results of electron biprism experiments are determined by the phase factor $U(2\pi)$, where $U(\theta)$ is given in (2.9). Also $U(2\pi)$ has the same value for the solutions ψ_1 and ψ_2. (This could not in general be true for $U(\theta)$ when $\theta \neq 0$ or $\theta \neq 2\pi$.)

2.6 Gauge Transformations and a Conjecture

For an electron in wave mechanics the proper gauge transformation is

$$\boldsymbol{A'} = \boldsymbol{A} + (\hbar/e)\operatorname{grad} f \quad , \qquad \psi' = \psi \exp(-\mathrm{i}f) \quad , \tag{2.16}$$

where f is a single-valued real function of $\boldsymbol{x}$ and is differentiable. The current density operator $\boldsymbol{j}$ (appearing just before (2.15b)) is invariant under the gauge transformation (2.16). Measurable quantities have to be gauge invariant in this way. Further discussion of the effect of this (ordinary) gauge invariance is given at the end of Appendix H.

Equations (2.15, 15a) would be given by (2.16) on using the function

$$f = (M - M')\theta \quad ; \tag{2.16a}$$

but here f is not single valued, so the flux transformation is not a gauge transformation. However (2.16a) provides a simple generalisation of a proper gauge transformation in the case that the space R (where $\boldsymbol{B} = 0$) has the topology of the region outside an infinite cylinder.

Since the expectation values in the space R of the physical operators $\Lambda_z, \boldsymbol{j}$ and the interference pattern phase factor $U(2\pi)$, are unaltered by any flux transformation, we conjecture that the wave functions ψ_1 and ψ_2 describe the same state in R. This is similar to the fact that ψ and ψ' in (2.16) describe the same state in a proper gauge transformation.

A consequence of this conjecture will be that in the case that the flux Φ in the solenoid is constant, the solutions ψ_1 and ψ_2 in (2.15a) can only give the same information about Φ. This information can only be the difference between Φ and an integral multiple of ϕ_e.

Outside the region R the flux transformation does not exist. Thus for $\rho < \rho_C$ in Fig. 2.1, we can use

$$A_\theta = \rho B/2 \quad , \qquad (0 \le \rho < \rho_C)$$

so $|A_\theta|$ does not vary as ρ^{-1}, and Λ_z cannot be defined independent of ρ as it was in (2.13). This prevents the existence of a flux transformation for $\rho < \rho_C$.

The flux transformation above is essentially the same as an invariance property proposed by T. T. Wu and C. N. Yang [2.4].

2.6.1 Notation for the Integral Part

For any positive or negative real number G we define $[G]$ to be the most positive (or least negative) integer g which obeys

$$g \le G \quad .$$

Then we write

$$G = [G] + \operatorname{rm} G \quad ,$$

so by definition

$$0 \le \operatorname{rm} G < 1 \quad .$$

2.6.2 Only Quantum Effects in R

Suppose that the flux $\Phi(t)$ in Fig. 2.1 is constant, or that it varies slowly with time, and assume that the confining potential $V(\rho)$ of Fig. 2.2 is present. Let the electron's wave function be ψ_1 in (2.15a).

Now choose an integer M' so that

$$M' - M = [P(t)] \quad . \tag{2.17}$$

It follows that the effective flux in the solenoid can be

$$\Phi'(t) = \phi_e \, \mathrm{rm} \, P(t) \quad ; \tag{2.17a}$$

so that

$$0 \leq \Phi'(t) < \phi_e \quad . \tag{2.17b}$$

Consequently we can think of the vector potential in R as being at most of order $O(h)$.

This is a satisfactory result. It is sometimes suggested that there is a large, and possibly detectable, vector potential in R. The use of the flux transformation is based upon the matrix elements and expectation values of physical operators. Within that limitation the detectable effects of the vector potential are $O(h)$.

2.7 Relation to Faraday's Induction

Suppose that electron wave packets circulate in the confining potential $V(\rho)$, as in the apparatus in Fig. 2.3. Suppose that the flux $\Phi(t)$ in the solenoid is now varied slowly. The classical physics description of what happens is as follows.

Choose the circulating electron to have angular momentum (about O_z) of amount $\Lambda < 0$ at time $t = 0$. Since $q = -e$ this implies $j_\theta > 0$. A current $j_\theta > 0$ is associated with a positive field intensity H_z, up through the ring; that gives a flux $\delta\Phi' > 0$ through the ring.

Lenz's law now shows that if the flux in the solenoid is altered by $\delta\Phi > 0$, then j_θ will be reduced and Λ will be altered by an amount $\delta\Lambda > 0$. As there is no resistance or dissipation in this ideal apparatus, the relation will be of the form

$$\delta\Lambda/\delta\Phi = \text{ a positive const. } . \tag{2.14a}$$

Strictly speaking we are here using Lenz's law in an average fashion over the cycle around the path within the confining potential.

We have to use more powerful methods to show that the constant in (2.14a) is independent of the geometry of the system.

Various authors have pointed out, directly or indirectly, that there is a deep connection between electromagnetic induction on the one hand and the

AB effect on the other. (A survey can be found in Ref. [2.5].) For example see: V. E. Weisskopf [2.6], E. Noerdlinger [2.7], M. Peshkin [2.8], F. Wilczek [2.9], D. H. Kobe [2.10]. Thus Weisskopf considers the work done on an electron in a fixed circular orbit as the magnetic induction is switched on.

The classical physics relation (2.14a) between the angular momentum component Λ and the induction flux Φ is the basic feature. We shall now find the constant on the right of (2.14a) by using a simple model.

2.7.1 Model for the Electromagnetic Induction Effect

We continue with classical methods. The Hamiltonian in (2.10) is not conservative since $P(t)$ varies. However if $d\Phi(t)/dt$ stays constant over a length of time, the problem can be related to a classical dynamical system.

Electromagnetic induction gives rise to an azimuthal electric intensity

$$E_\theta = -\dot{A}_\theta \ .$$

In our model this does not alter with time. In place of A, an electric potential $V(\theta)$ can be used, with

$$E_\theta = -\rho^{-1}\partial V(\theta)/\partial\theta \ .$$

By (2.3)

$$\partial V(\theta)/\partial\theta = (2\pi)^{-1}\dot{\Phi} \ . \tag{2.18}$$

Integrating gives

$$V(\theta) = (2\pi)^{-1}\theta\dot{\Phi} \ ,$$

and θ must be allowed to run from $-\infty$ to $+\infty$. This gives a *ramping force* that will drive the charge around the ring, so long as $\dot{\Phi}$ is constant. It corresponds to a ramping potential energy

$$V_{\mathrm{ra}} = qV(\theta) \ , \tag{2.18a}$$

where, for generality, q is the charge of the particle.

Thus the Hamiltonian for the model is

$$H_{\mathrm{M}} = p^2/2m + V(\rho) + V_{\mathrm{ra}}(\theta) \ , \tag{2.19}$$

where m is the mass of the particle and $V(\rho)$ is the confining potential. Analogous to the canonical equation

$$\dot{p}_x = -\partial H/\partial x$$

we have

$$\dot{L}_\theta = -\partial H_{\mathrm{M}}/\partial\theta = -\partial V_{\mathrm{ra}}(\theta)/\partial\theta \ , \tag{2.19a}$$

L_θ being the angular momentum about O_z. From (2.18, 18a)

$$\dot{L}_\theta = -(q/2\pi)\dot{\Phi} \ . \tag{2.20}$$

Thus L_θ is the same quantity as Λ in (2.14a). That equation becomes

$$\delta \Lambda = \frac{e}{2\pi}\delta \Phi \ . \tag{2.20a}$$

Notice that the proof actually does not require the path of the electron in Fig. 2.3 to be a circle, but it does require the confining potential $V(\rho)$ to be independent of θ (i. e. the confining forces are central forces).

2.8 Quantum Treatment, Bloch's Theorem and Acceleration

In Sect. 2.7 the relation in (2.20a) between the change in flux $\delta \Phi$ through an orbit and the change in angular momentum $\delta \Lambda$ was derived classically. The wave mechanical treatment of the same physical problem leads us via Bloch wave functions to a simple form of the AB result.

It is desirable to give some equivalent wave mechanical relation to replace (2.20a). The method used in Sect. 2.7 involves the angle θ, and there are difficulties in using an angle as a quantum variable. These difficulties are examined in [2.11]. A later work by Barnett and Pegg [2.12] puts forward a method of avoiding these problems, but for reasons of compactness that method will not be used here.

Instead of the angle method which could start from (2.18), the coupling (p, A) will be used. This has the interesting aspect that it links up with the theory of electron motion in a crystal, and also with the concept of acceleration in a quantum system. The method moreover emphasises the basic importance of the propagation vector k, rather than the electron's velocity, which is the group velocity v.

Consider Schrödinger's equation for an electron in a potential $V(x)$ that is periodic in ordinary three-dimensional space. The basic unit of periodicity is a cell that we take to be a rectangular parallelepiped and its periodicity extends to infinity in each dimension. The equation is

$$H\psi \equiv \left(p^2/2m + V(x)\right)\psi(x) = E\psi(x) \ , \tag{2.21}$$

where E is the eigenvalue of energy and

$$p \equiv \frac{\hbar}{\mathrm{i}}\nabla$$

is the momentum operator. A basic result is that any bounded solution ψ of (2.21) can be written in the form

$$\psi(x) = \exp\left(\mathrm{i}k \cdot x\right) u_k(x) \tag{2.22}$$

where k is a real vector and the function $u_k(x)$ is periodic with the same basic cell. $\psi(x)$ is, in general, not periodic. It is a modulated plane wave, having periodic modulation. The kinetic energy of the electron will vary over the cell since $V(x)$ varies, but the propagation vector k is constant. k will be used as a quantum number to label the solution in (2.22).

The modulating function obeys the equation

$$\{(p + \hbar k)^2/2m + V(x)\}u_k(x) = Eu_k(x) \ . \tag{2.23}$$

These results constitute Bloch's theorem [2.13] for the motion of an electron in a crystal; Eq. (2.22) gives Bloch's wave function. The idea goes back to Floquet [2.14]. A precise account of Bloch wave functions is found in [2.15] and [2.16].

The spectrum of values of E will, in general, not be continuous. There are gaps separating bands of energy eigenvalues. This is, of course, important in electrical conduction phenomena.

The use of instantons in deriving quantum tunnelling in a dynamical system having degenerate classical ground states, such as these conductors, is demonstrated in [2.17]

2.8.1 Constant Acceleration

Interesting results follow when an electric field E, constant in space and time, is applied to the above system. The situation is described by the potential

$$A = -Et \tag{2.24}$$

and the Hamiltonian is given by (2.4). Thus Schrödinger's equation is

$$\{(p - eEt)^2/2m + V(x)\}\psi(x) = E\psi(x) \ . \tag{2.25}$$

We write

$$\psi(x) = \exp(ieE \cdot xt/\hbar)\psi_0(x) \ , \tag{2.26}$$

where $\psi_0(x)$ will obey (2.21).

Since $V(x)$ is periodic, Bloch's theorem applies to $\psi_0(x)$; so for any real k in an allowed band of propagation vectors k, a solution is

$$\psi_0(x) = \exp(ik \cdot x)u_k(x) \ . \tag{2.27}$$

Notice that $u_k(x)$ is a periodic modulating factor that does not depend on E or t. Hence

$$\psi(x) = \exp[i(k \cdot x + eE \cdot xt/\hbar)]u_k(x) \tag{2.28}$$

is a solution of (2.25). (In this analysis we can regard t merely as a parameter in H.)

In (2.28) we can write

$$k' = k + eEt/\hbar \tag{2.29}$$

for the effective propagation veector k'. Since $u_k(x)$ is independent of E or t, the vector k is an index, or a quantum number, for the solutions. The wave function $\psi_0(x)$ gives the solutions at $t = 0$. In order to keep to the same solution as time develops, Eqs. (2.28) and (2.29) show that k should be varied according to the rule

$$\hbar \dot{k} = -eE \tag{2.30}$$

Classically this means that the electron is steadily slowed down by the field E.

2.8.2 What Determines the Solution?

The necessity for the latter result is as follows. Equation (2.25) is solved at a sequence of values $t_1, t_2, t_3, \ldots$ of t, giving the solutions

$$\psi(\boldsymbol{x}, t_j) \quad , \qquad j = 1, 2, 3, \ldots$$

and each of them is of the form in (2.28). Continuity in t is necessary. Thus if the number of values t_j is increased so that the separation

$$|t_j - t_{j+1}|$$

between neighbours becomes small, the condition

$$|\psi(\boldsymbol{x}, t_j) - \psi(\boldsymbol{x}, t_{j+1})| < \varepsilon \quad , \qquad \text{for all } x \quad , \tag{2.28a}$$

has to be obeyed where ε is a small number. We require this to hold for all x since the crystal is infinite in extent.

On letting $\varepsilon \to 0$, Eq. (2.28a) becomes the condition for $\psi(\boldsymbol{x}, t)$ to be continuous in t, uniformly over all $\boldsymbol{x}$.

Now looking at the exponential factor

$$\exp(\mathrm{i}\boldsymbol{k'} \cdot \boldsymbol{x})$$

in $\psi(\boldsymbol{x})$, Eq. (2.28), it is clear that to satisfy (2.28a) we require $\boldsymbol{k'}$ to be independent of t. As t is varied, $\boldsymbol{k}$ will now alter slowly and the modulating factor $u_{\boldsymbol{k}}(\boldsymbol{x})$ will alter slowly and smoothly within the energy band of values of $\boldsymbol{k}$. Equation (2.30) follows.

It is also clear that the relation of an energy level to its neighbours will not be altered due to the presence of the field $\boldsymbol{E}$, so the energy bands will remain. It is necessary that $|\boldsymbol{E}|$ is sufficiently small that (2.25) remains valid for the duration of any measurement. Also $|\boldsymbol{E}|$ must be sufficiently small that transitions from one band to another are negligible. This is a form of adiabatic condition.

Equation (2.30) is a well known result in the theory of crystalline conductors. It is remarkable in that Newton's law has a different left-hand side, namely it has $m\dot{v}$. A discussion of the physics of this feature can be found, for example, in Chap. 11 of Ref. [2.16].

2.8.3 The Change in Phase and the AB Effect

Let δX denote the change in a physical quantity X in the time δt. Then by (2.30)

$$\hbar \delta \boldsymbol{k} = -e\boldsymbol{E}\delta t = e\boldsymbol{A} \quad . \tag{2.31}$$

Let $\boldsymbol{k}$ and $\boldsymbol{E}$ lie along an edge $\boldsymbol{a}$ of the basic cell. Then the *change* in the phase difference of the electron wave between the ends of $\boldsymbol{a}$, in time δt, is

$$\delta \int_{(a)} \frac{2\pi}{\lambda} \mathrm{d}s = \frac{e}{\hbar} \delta \int_{(a)} \boldsymbol{A} \cdot \mathrm{d}s \quad . \tag{2.32}$$

Here (a) denotes that the integration over ds or ds, is along the cell edge a. A is constant in space and its magnitude alters linearly with time. Equation (2.32) is closely related to the AB effect.

In the case of a metal, a three-dimensional basic cell of atomic size is used. Here we shall go over to a one dimensional cell of macroscopic size. Consider the electron moving around the solenoid in a closed confining potential, as indicated in Figs. 2.3 and 2.4. The basic cell is one passage around the solenoid in the confining potential. Motion in the trough of the potential gives a band of closely spaced energy levels. Tranverse vibrations give rise to a set of such bands.

The confining potential $V_C(x)$ need not be of the same depth at different positions around the closed circuit. Also the circuit need not be a circle, but it should be a smooth closed curve. If the wavelength λ is very much less than the typical radius of curvature R, we shall assume that a one-dimensional form of (2.21) can be used.

Also we assume that an electric field E can be set up that is everywhere tangential to the lowest line in $V_C(x)$, and that $|E|$ is constant along the ring, and in time. It is assumed that A obeys (2.24). In this case (2.32) will be valid in the form

$$\delta\eta(k) = \frac{e}{\hbar}\delta\Phi \tag{2.33}$$

where $\eta(k)$ is the phase *difference* of the electron wave on going once around the solenoid, and Φ is the flux through the orbit. δ has the same meaning as in (2.31).

Equation (2.33) is the wave mechanical form of (2.20a) in Sect. 2.7 for the electromagnetic induction effect.

Consider the case in which the confined track is a circle whose centre is on the solenoid. The wave function, in the notation of Sects. 2.4, 2.7, will contain the factor

$$\exp(i\Lambda\theta/\hbar) \ .$$

In this case $\eta = 2\pi\Lambda/\hbar$, and (2.33) yields

$$\delta\Lambda = (e/2\pi)\delta\Phi \ ,$$

in agreement with the classical calculation leading to (2.20a) in Sect. 2.7.

2.9 Composition of the Angular Momentum in R

The important classical result in (2.20a) is used to interpret (2.13) or (2.14), for the angular momentum of a wave packet moving in the field free space R – under the influence of the confining potential $V(\rho)$ – around the solenoid, as in Fig. 2.3. Λ_z is the sum of the angular momentum operator L_z plus the angular momentum $(e/2\pi)\Phi$ that is induced on raising the flux from zero to Φ. Thus the *base point* for the eigenvalue Λ is the zero eigenvalue for L_z

at flux $\Phi = 0$. This comment is the easiest way to understand the solenoid problem.

In our analysis it has been assumed that $\Phi(t)$ alters sufficiently adiabatically. In (2.20a) a change of flux by ϕ_e gives a change $\hbar$ in angular momentum. $\Phi(t)$ should only change by an amount ϕ_e over several, or more, turns of the wave packet around the solenoid – cf. (2.10b). A change that is so slow in terms of θ will only give a small matrix element between states whose eigenvalues for L_z differ by $\pm\hbar$. Thus the term $M\hbar$ in (2.14) should be unaltered by the slow change in $\Phi(t)$. (Such adiabatic quantities are also discussed in the classical treatment in Sect. 4.4 below.)

In (2.14) there is some overlap between $M\hbar$ and $(e/2\pi)\Phi$. In R, we can, without any physical effect, add (or subtract) units of ϕ_e to (or from) Φ, so long as

$$\hbar(M + [\Phi/\phi_e])$$

is unaltered.

The interesting part of Λ is $\delta\Lambda'$ which obeys

$$0 \leq \delta\Lambda' < \hbar \ . \tag{2.34}$$

It correspondens to $\Phi'(t)$ of (2.17b). This non-integral part of $\Lambda/\hbar$ cannot be avoided unless there is a law that the flux Φ can only appear in integral multiples of ϕ_e. Experiments show that this is not true in general. Superconducting quantum interference devices (SQUID) or similar devices, can easily be made to measure flux more precisely than ϕ_{2e}, where $\phi_{2e} = (1/2)\phi_e$. Flux variations as small as $10^{-5}\phi_{2e}$ have been measured (see Refs. [2.18]–[2.20]).

It should be noted that SQUIDs do not absolutely measure very small fluxes; but they can measure very small *changes* in flux on using a compensation device. That is enough for our argument here. Further discussion on superconductor measurements will be found in Sect. 5.11 and in Chap. 6.

A method that might be used to measure the non-integral part of $\Lambda/\hbar$ is discussed below, Sect. 2.10. The question of the limit on the accuracy of flux measurements is discussed in Chap. 6.

2.10 Nature of the Wave Functions

Consider the same motion around the cylindrical solenoid as in Fig. 2.3. Equation (2.6) gives the relation between the wave function $\psi_A(x)$ for this setting of the apparatus and a solution $\psi_0(x)$ of (2.5a).

If $\psi_A(x)$ is a single-valued wave function, then $\psi_0(x)$ cannot have that property, unless Φ/ϕ_e is an integer. The solution ψ_A is of the form in (2.11), and since the factor $\chi(\rho,t)$ is common to ψ_A and ψ_0, we omit it, and write

$$\psi_A = \exp(iM\theta) \ . \tag{2.35}$$

Using (2.6) and (2.9)

$$\psi_0 = (U(\theta))^{-1}\psi_A$$
$$= \exp\{i\,(M''+r)\,\theta\} \tag{2.35a}$$

where

$$M'' = M + M' \ , \qquad M' = [\Phi/\phi_e] \ , \qquad r = \mathrm{rm}(\Phi/\phi_e) \ .$$

By definition $0 \le r < 1$, and we assume $r \ne 0$; therefore

$$\psi_0(\theta = 2\pi) \ne \psi_0(\theta = 0) \ .$$

The AB relation does not in general relate two single-valued wave functions for different settings of the apparatus. It does not in general relate two eigenstates.

However, in dealing with electro-optical interference the AB relation is relevant, because (2.6) can give the electro-optical path difference between fixed points for two different settings of the apparatus. Such wave packets are not eigenstates.

2.10.1 Expansion of $\psi_0(x)$ and Measurement of $\mathrm{rm}(\Lambda/\hbar)$

The above shows that $\psi_0(x)$ is not in general an eigenstate of the Hamiltonian for movement in the confining potential $V(\rho)$ for the setting $\Phi = 0$. However, it can be expanded in a series of such eigenstates. This will show how non-integral $(\Lambda/\hbar)$ gives rise to a superposition of these states.

On ignoring the factor χ, the $\Phi = 0$ eigenstates are $\exp(in\theta)$, with n an integer. For $0 < r < 1$ there is a summation:

$$\lim_{N\to\infty} \sum_{-N}^{+N} \frac{\exp(in\theta)}{n-r} = -\frac{\pi}{\sin \pi r}\exp\{ir(\theta-\pi)\} \ , \qquad \text{for } 0 \le \theta \le 2\pi \ .$$

Outside $(0, 2\pi)$ the value of the sum repeats with the period 2π. Using (2.35a)

$$\psi_0(x) = \sum_{-\infty}^{\infty} a_n \exp(in\theta) \ , \qquad (0 \le \theta \le 2\pi) \tag{2.36}$$

with

$$a_n = \frac{\sin \pi r}{\pi}\exp(inr)\frac{1}{M''+r-n} \ . \tag{2.36a}$$

For $r = 0$ only the one coefficient $a_{M''}$ (with $M'' = n$) in the series in (2.36) is non-zero. For $r = 1/2$ the central coefficients, in order, are

$$\ldots, 2i/3\pi, 2i/\pi, -2i/\pi, -2i/3\pi, \ldots \ .$$

The signature for $\Lambda/\hbar$ not being an integer is that there is a number of non-zero coefficients, and there is a change of phase by $\exp(i\pi)$ at the largest one.

3. Generalization
of the Aharonov–Bohm Effect

It is shown how the AB effect can be seen in other phenomena in which a charged particle moves in a magnetic field. The preliminary steps are setting up a simple quantum theory solution for motion in a uniform induction field and in a special family of axially symmetric fields.

3.1 Motion in a Uniform Induction B

There is considerable similarity between the motion around a solenoid and a solution for non-relativistic motion in a uniform induction.

In the solenoid case the flux Φ through the orbit is constant and in the case of the uniform field Φ is the same for all orbits having the same energy. In both cases the component Λ_z of angular momentum about the symmetry axis Oz is the sum of a quantum part L_z and a flux part, as in (2.13) above. In both cases a wave function ψ of the form in (2.11) is used. It should be noted that this form of wave function does not in itself restrain the motion to be circular.

The uniform field is $(0, 0, B)$, $B > 0$, in Cartesian coordinates. Simplicity depends on a suitable choice of gauge. We use

$$\boldsymbol{A} = (0, 0, \rho B/2) \tag{3.1}$$

in cylindrical coordinates (z, ρ, θ). The Hamiltionian of (2.4) becomes

$$H = -\frac{\hbar^2}{2m}\left[\frac{\partial^2}{\partial z^2} + \rho^{-1}\frac{\partial}{\partial\rho}\left(\rho\frac{\partial}{\partial\rho}\right)\right] + \frac{1}{2m}\left(\rho^{-1}L_z + e\rho B/2\right)^2 \tag{3.2}$$

with L_z defined in (2.13a). We write

$$L_z + eB\rho^2/2 = \hbar\left(\frac{\partial}{i\partial\theta} + \Phi(\rho)/\phi_e\right) \,, \tag{3.2a}$$

where

$$\Phi(\rho) = \pi\rho^2 B \tag{3.2b}$$

is the flux through any circle of radius ρ lying in the Oxy plane. The origin of coordinates is arbitrary, and we choose it to be at the centre of the circle on which the wave packet lies.

The form

$$\psi = \chi(\rho, t) \exp(iM\theta) \ , \qquad (M = \text{integer}) \tag{3.3}$$

is used to solve the wave equation

$$i\hbar \frac{\partial \psi}{\partial t} = -\left\{ \frac{\hbar^2}{2m} \left[\frac{\partial^2}{\partial z^2} + \rho^{-1} \frac{\partial}{\partial \rho} \left(\rho \frac{\partial}{\partial \rho} \right) \right] \right.$$

$$\left. - \frac{1}{2m} \left(\rho^{-1} L_z + e\rho B/2 \right)^2 \right\} \psi \ . \tag{3.3a}$$

There is no motion along Oz in (3.3), but a broad wave packet along Oz could easily be used.

Choose a positive integer M, and define a positive number ρ_0 by the relation

$$M\hbar = eB\rho_0^2/2 \ ; \tag{3.4}$$

in other words,

$$M = \Phi(\rho_0)/\phi_e \ . \tag{3.4a}$$

It follows that

$$(\rho^{-1} L_z + eB\rho/2)\psi = \{eB\rho_0 + eB(\rho - \rho_0)^2/2\rho\} \psi \ . \tag{3.5}$$

This is the basic relation for cyclotron motion. We shall use the quantity ω_B defined by

$$\omega_B = eB/m \ . \tag{3.5a}$$

It will be recognised as the cyclotron frequency of classical theory.

The Method. In order to find the main features of the motion a simple form of the variational method will be used here. It is known that in general the variational method is better in determining eigenvalues than in finding wave functions, and this also becomes clear here. An exact wave-mechanical solution is not trivial; one is given in Sects. H.2, H.3 in Appendix H, and the Landau–Lifschitz method is demonstrated in Sect. H.1.

3.2 Cyclotron Motion

For a first approximation we shall ignore the terms in $\partial/\partial\rho$ and $\partial^2/\partial\rho^2$ (as well as $\partial^2/\partial z^2$) in (3.3a). In the state ψ of (3.3) the Hamiltonian has the expectation value

$$\frac{\pi(eB)^2}{m} \int_0^\infty \left[\rho_0 + (\rho - \rho_0)^2/2\rho \right]^2 |\chi(\rho)|^2 \rho \, d\rho \ .$$

The normalization is

$$2\pi \int_0^\infty |\chi(\rho)|^2 \rho \, d\rho = 1 \ .$$

The expectation value is a minimum

$$T_\theta = \frac{(eB\rho_0)^2}{2m} \quad , \tag{3.6}$$

for some wave function $\chi(\rho)$ that is very small except for ρ near to ρ_0.

The relation between the extrema of expectation values of Hermitian operators and their eigenstates shows that we have the approximate solution, and T_θ is the kinetic energy of the circular motion.

From (3.4, 5a) it follows that

$$T_\theta = M\hbar\omega_B \quad . \tag{3.6a}$$

This last relation, like (3.4), depends upon B_z being uniform inside, as well as on, the circular orbit; it need not be true otherwise. This fact is an indication of a connection with the AB effect.

It may be useful to notice that in practice M is a large integer, since it is of the order of the radius of the orbit divided by λ.

The wave function is of the form

$$\psi = \psi(\rho) \exp\left\{i(L'_z\theta - T_\theta t)/\hbar\right\} \quad , \tag{3.7}$$

where L'_z is the eigenvalue of L_z. A circulating wave packet can be made from the sum of such terms having adjacent values of M. The phase, and group, angular velocities

$$T_\theta/L'_z \quad \text{and} \quad \frac{\delta T_\theta}{\delta M} \bigg/ \frac{\delta L'_z}{\delta M}$$

respectively, both equal ω_B.

However if we wish to use the classical relation

$$\omega_B = (\text{angular momentum})/m\rho_0^2 \quad ,$$

it is necessary to take the angular momentum operator to be

$$\Lambda_z = L_z + eB\rho^2/2 \quad . \tag{3.8}$$

Thus the eigenvalues for state ψ above is

$$\Lambda = M\hbar + eB\rho_0^2/2 \quad . \tag{3.8a}$$

Equation (3.4) now gives

$$\Lambda = eB\rho_0^2 \quad , \tag{3.8b}$$

and that gives the correct value for ω_B, as in (3.5a). Λ_z is just the operator in (3.2a). Here, as in the case of the solenoid (in Eq. (2.13)), Λ_z is the sum of L_z and a classical term.

Notice that in this first approximation the spread of $|\chi(\rho)|$ about the mean value ρ_0 is assumed to be small, and the use of ρ_0^2 in the last term of (3.8a) is valid. On using the next approximation it is clear from (3.9c) that $(2n+1)/4M$ should be small for this simple interpretation to remain valid.

3.2.1 Orbit Quantization: The de Haas–van Alphen Effect

Equation (3.8b) shows that the peripheral momentum p_θ equals $eB\rho_0$. Using $p_\theta = \hbar k$, it is clear that (3.4) quantizes the space of free electron propagation vectors with circles of area

$$\pi k^2 = \frac{2\pi}{\hbar} eBM \ , \qquad (M = \text{integer}) \tag{3.4b}$$

the circles being orthogonal to $\boldsymbol{B}$. This result has important consequences in solid state physics. (The precise form of (3.4b) depends on the induction being a uniform field $\boldsymbol{B}$.)

The area in $\boldsymbol{k}$-space (perpendicular to $\boldsymbol{B}$) of quantization of propagation vectors, as in (3.4b), is of basic importance for understanding the de Haas–van Alphen effect. That effect is the periodic variation of the magnetic suscepti-bility of a pure metal as B is varied (at suitable temperatures). The periodic effect is in the variable B^{-1}. There are also other similar effects as B is var-ied, such as an effect in the electrical resistivity (the Schubnikow–de Haas effect). For some account of the theory of these phenomena see Sects. 9.6, 9.7 of Ref. [3.1], and Chap. 11 of Ref. [3.2].

Consider an area of cross-section of the Fermi surface (with fixed k_z and B along Oz). As B is varied, by (3.4b) the energy levels vary. As a result the free energy of the electron gas oscillates; the period of oscillation corresponds to the passage of successive quantized orbits through the Fermi level.

3.3 Next Approximation and the Radial Oscillations

The following is a quick but imprecise method; an accurate treatment is to be found in Sects. H.2, H.3 of Appendix H.

Apart from $\partial^2/\partial z^2$, the remaining terms in the Hamiltonian in (3.3a) are

$$-\frac{\hbar^2}{2m}\rho^{-1}\frac{\partial}{\partial\rho}\left(\rho\frac{\partial}{\partial\rho}\right) + \frac{(eB)^2}{2m}(\rho - \rho_0)^2 + (\text{small terms}) \ . \tag{3.9}$$

Let b be the width in the ρ-direction of the wave function. On assuming that

$$b/\rho_0 \ll 1 \ , \tag{3.9a}$$

the $\partial/\partial\rho$ term in (3.9) can be omitted. Ignoring the "small terms" in (3.9), we are left with the Hamiltonian for harmonic oscillations in ρ about ρ_0 with (circular) frequency ω_B. The energy levels are

$$E_n = \left(n + \tfrac{1}{2}\right)\hbar\omega_B \ ; \qquad (n = 0, 1, 2, \ldots) \tag{3.9b}$$

the corresponding variances are

$$\left\langle (\rho - \rho_0)^2 \right\rangle_{(n)} = \frac{n + \tfrac{1}{2}}{2M}\rho_0^2 = \left(n + \tfrac{1}{2}\right)\frac{\phi_e}{2\pi B} \ . \tag{3.9c}$$

The term in $(\rho - \rho_0)^2$ is not a good approximation in (3.9) above, except when $|\rho - \rho_0|$ is small. Equation (3.9c) suggests that for cases with $n \ll 4M$, terms in $(\rho - \rho_0)^4$ can be omitted from (3.9). For larger values of n there are appreciable corrections to (3.9c). The exact result is discussed at the end of Sect. H.3. The exact solution in Appendix H also shows that the energy eigenvalue in (3.9b) is correct for motion in a uniform induction $\mathbf{B}$. [The values in (3.9b,c) would be the correct values for a one-dimensional simple harmonic oscillator with frequency ω_B.]

The zero-point energy (for $n = 0$) shows that the wave function $\chi(\rho)$ has a finite spread about ρ_0, and thus it improves the first approximation above.

The total energy is

$$E = \left(M + n + \tfrac{1}{2}\right) \hbar \omega_\mathrm{B} \ . \tag{3.10}$$

The quantum numbers M and n relate to the azimuthal and the radial motion respectively. Increasing M increases the radius ρ_0 of the orbit, and increasing n increases the spread of the wave function about ρ_0, as in (3.9c). The corresponding wave functions are obvious.

Comments

(i) The following are classical results: the values for ω_B in (3.5a), for Λ in (3.8b), and for T_θ in (3.6).

(ii) An important difference between the solenoid problem and the uniform field problem is that in (2.14) integral multiples of $\hbar$ can be shifted between $M\hbar$ and $(e/2\pi)\Phi$ without causing any physical effect. That freedom does not exist in motion under the uniform field, since (3.4a) defines the flux through the mean radius ρ_0 of the wave packet in terms of M.

(iii) *The Gauge.* The role of the gauge is important. Here $\mathbf{A}$ has been chosen as in (3.1), but in Appendix H, for the same physical problem, the gauge $\mathbf{A} = (-By, 0, 0)$ is used. Different gauges may give different wave functions $\psi(\mathbf{x})$ for the same problem, but they should yield the same gauge-invariant operators, or expectation values, such as the current density $\mathbf{j}(\mathbf{x})$. $\mathbf{j}(\mathbf{x})$ is defined just before (2.15b) and in (H.5). A further example of gauge invariance can be seen at the end of Sect. H.3 and also in the discussion of the physical interpretation of the result of Sect. H.1. (See also pp. 456–458 in Ref. [3.3].)

It should be remembered that the simple form of the wave function ψ in (3.3) is useful because the induction field $\mathbf{B}$ is such that $\mathbf{A}$ can be of the form $(0, 0, A_\theta(\rho))$ (in cylindrical coordinates). Thus $\mathbf{B}$ is restricted to the form $(B_z(\rho), 0, 0)$.

(iv) *Ginsburg–Landau Theory.* The results in (3.9b,c) appear in another form in the Ginsburg–Landau (GL) macroscopic theory of BCS supercouductors [3.4], [3.5]. It is of interest to give a brief account of how the GL theory is related to the solutions of (3.9).

In the GL theory the order parameter is taken to be the macroscopic wave function $\psi(\mathbf{x})$. The density of free-energy is assumed to have the form:

$$F(\mathbf{x}) = F_n + \alpha|\psi(\mathbf{x})|^2 + (\beta/2)|\psi(\mathbf{x})|^4 + (\hbar^2/2m)|\nabla\psi(\mathbf{x})|^2 \ . \tag{3.11}$$

The parameters F_n, α, β may depend on the temperature, and in the GL theory m (in the last term) is taken to be the mass of the electron, although ψ is the wave function of the Cooper pair. Here $|\psi|^2$ can be normalised as n_s, the superconducting electron density. (And $n_s = 2n_C$ where n_C is the density of Cooper pairs.) (For further discussion see Ref. [3.6].)

The last term on the right of (3.11) gives stiffness to the macroscopic wave function $\psi(x)$, as can be seen from the need to keep down the free energy. Strong variations in $\psi(x)$ that are smaller in extent than the length $f(T)$, where

$$f(T) = \left(\frac{\hbar^2}{2m|\alpha|}\right)^{1/2} \quad , \tag{3.12}$$

are suppressed. $f(T)$ is a coherence length for $\psi(x)$ and it is a function of the parameter α.

In the GL theory for type II superconductors there is a critical magnetic intensity H_{C2} above which the order parameter $\psi(x)$ vanishes and there is no superconductivity. Near this transition the higher order term in $|\psi(x)|^4$ in (3.11) can be ignored. Minimising the Gibbs free energy for the problem now gives the equation;

$$\frac{1}{2m}\left(\frac{\hbar}{i}\nabla + 2eA\right)^2 \psi(x) + \alpha\psi(x) = 0 \tag{3.13}$$

for the behaviour of $\psi(x)$ in the presence of the magnetic field.

Comparing solutions of (3.13) with the analogous solutions for the electron at the start of Sect. 3.3 above, it follows from (3.9c) that the least value of

$$\langle(\rho - \rho_0)^2\rangle$$

occurs for $n = 0$. The same will be true for $(f(T))^2$ in connection with (3.13). For that case

$$|\alpha| = \hbar\omega_C/2 \quad , \tag{3.13a}$$

where $\omega_C = 2\omega_B$.

Equation (3.9c) also shows that a non-zero value of $f(T)$ implies that the field H cannot exceed a maximum value. In this model that is the critical field H_{C2}. By (3.12, 13a) it is given by

$$\mu_0 H_{C2} = \frac{\hbar}{qf^2(T)} = \frac{\phi_q}{2\pi f^2(T)} \tag{3.14}$$

(where $q = 2e$).

Finally, using (3.9c), the smallest variance is

$$\langle(\rho - \rho_0)^2\rangle_{\min} = \frac{\hbar}{2q\mu_0 H_{C2}} = \frac{f^2(T)}{2} \quad . \tag{3.15}$$

This agrees with the idea of the parameter $f(T)$ as expressed above. (See for example, Chap. 8 in [3.7] for a detailed discussion of the GL theory and its applications.)

3.4 The "Field-Free" Wave Function (Uniform B_z)

In order to see the basic nature of the AB relation of (2.6, 7), we shall generalize the relation to cases other than the solenoid. Motion in a uniform induction B, as discussed in the preceding sections, is useful for this purpose.

In cylindrical coordinates the vector potential in (3.1) is

$$A = (0, 0, A_\theta) \quad , \qquad A_\theta = \rho B/2 \quad . \tag{3.16}$$

The function $U_A(x)$ of (2.7) is undefined in a region where B is not zero; it is necessary to specify the *integration path* from x_0 to x.

When that is done we shall write the function as $\bar{U}_A(x)$.

Assume that the electron wave packet moves around a circle $\rho = \rho_0$ lying in the plane $z = 0$. The wave packet will have a small radial spread about this circle. The minimum value of radial spread, as we saw in Sect. 3.3, is given by (3.9c) for the lowest state $(n = 0)$ of radial oscillation.

We have

$$\rho_0^{-1} \left(\langle (\rho - \rho_2)^2 \rangle_{(0)} \right)^{1/2} = \frac{1}{2M^{1/2}} = \frac{1}{2} \left(\frac{\phi_e}{\Phi(\rho_0)} \right)^{1/2} \tag{3.9d}$$

where (3.4a) is used. The integer M obeys $M \geq 1$, but (3.9d) shows that we require M to be fairly large in order to give a well-defined orbit even when the lowest value, $n = 0$, is used. The result in (3.9d) is to be multiplied by $(2n + 1)^{1/2}$ for $n > 0$.

In place of $U_A(x)$ of (2.7), we define

$$\bar{U}_A(\theta) = \exp \left\{ -(\text{i}e/2\hbar)B\rho_0^2\theta \right\} \quad . \tag{3.17}$$

The integration has been taken along the path of the wave packet, starting at $\theta = 0$. For a complete circuit this gives

$$\bar{U}_A(2\pi) = \exp \left\{ -2\pi \text{i} \Phi(\rho_0)/\phi_e \right\} \quad , \tag{3.17a}$$

where $\Phi(\rho)$ is defined in (3.2b).

Now consider the formulae in (2.35, 35a). For the moment the factor $\chi(\rho, t)$ in the wave function ψ of (3.3) will be ignored. We write ψ as

$$\psi_A(\theta) = \exp(\text{i}M\theta) \quad , \qquad (M = \text{integer}) \quad . \tag{3.18}$$

Analogous to (2.35a) we define a new wave function:

$$\psi_0(\theta) = \left(\bar{U}_A(\theta) \right)^{-1} \psi_A(\theta)$$
$$= \exp \left[\text{i}\theta \left(M + eB\rho_0^2/2\hbar \right) \right] \quad . \tag{3.18a}$$

This procedure of going from (3.18) to (3.18a) is just copied from the solenoid case on using $\bar{U}_A(\theta)$ of (3.17). The result however is

$$\psi_0(\theta) = \exp(\text{i}\varLambda\theta/\hbar) \tag{3.18b}$$

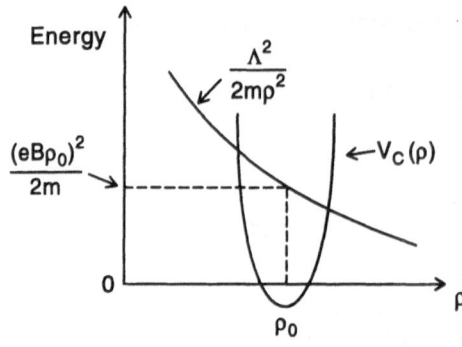

Fig. 3.1. Centrifugal potential $\Lambda^2/2m\rho^2$ and the confining potential $V_C(\rho)$

where Λ is the (total) angular momentum as given by (3.8a). Thus $\psi_0(\theta)$ describes the motion of the electron on the circle of radius ρ_0, having kinetic energy T_θ given in (3.6).

The Lorentz force in (2.1) does not alter the energy, so $\psi_0(\theta)$ would be the solution of Schrödinger's equation, Eq. (2.4a), for an electron confined by a potential $V_C(\theta)$ alone, so that it moves on the circle $\rho = \rho_0$ with angular momentum Λ. Thus (3.18a) is a generalization of the AB relation. We call $\psi_0(x)$ the "field-free" wave function.

In Fig. 3.1 the confining potential and the centrifugal potential are shown. $\chi_0(\rho)$, the radial component of ψ_0, will be negligible except in a small segment around the minimum of

$$\frac{\Lambda^2}{2m\rho^2} + V_C(\rho) \ .$$

The minimum is to be at ρ_0. The minimum value of $V_C(\rho)$ itself is not zero, but is put at the value

$$V_C(\rho_0) = \tfrac{1}{2}(\omega_B - \omega_C) \ . \tag{3.18c}$$

ω_B is given in (3.5a), and ω_C is the corresponding frequency in this model. The condition (3.18c) ensures that in the lowest state ($n = 0$) the energy of the model in Fig. 3.1 and that in the uniform field B are equal.

In general the factors $\chi_0(\rho)$, and $\chi(\rho)$ of (3.3) will differ somewhat, and higher levels of radial oscillation ($n \geq 1$) need not agree in the two cases. Since it is the motion along the θ-direction that is of importance here, any other difference in χ between ψ_A and ψ_0 will be ignored in our discussion.

For the uniform field B the generalized AB phase given by (3.17) is half of the phase change in ψ_0 on passing round the arc θ of the circle $\rho = \rho_0$. The basic origin of this factor $\tfrac{1}{2}$ will be seen below in discussing the classical physics picture. In practice the phase of $U_A(2\pi)$ may be large, since M is of the order of $2\pi\rho_0/\lambda$. It is large on account of the large area of flux enclosed by the orbit. In the solenoid case, in practical interference experiments, the corresponding area is very small (cf. Sect. 5.1 below).

3.4.1 The AB Effect in Disordered Conductors

It was seen in Sect. 3.2 that the quantization of electron orbits in a uniform induction B gives rise to the de Haas–van Alphen effect (c.f. Sects. 9.6, 9.7 Ref. [3.1] and Chap. 11 of Ref. [3.2]). This is a periodic effect in magnetic susceptibility as B is varied. A similar periodic behaviour in the electrical resistivity is the Schubnikov–de Haas effect [3.1], [3.2].

Quite a different type of effect arises from the AB effect in the case of two-dimensional flow of current in a disordered conductor. The phenomenon is known as magneto-resistance.

Electron wave packets can move forwards and backwards along trajectories which surround a hole in the conductor. An important set are the self-crossing trajectories. A simple example is wave packets on a circle. We choose a point O on the circle. The phase factor at the centre of a forward-going wave packet is

$$\exp(ik\theta) \ , \qquad (k = \text{const.})$$

θ being the angle measured from O. At the centre of the backward-going wave packet the phase factor is

$$\exp(-ik\theta) \ , \qquad (k = \text{const.}) \ .$$

In the first case the difference in phase factor on going once around the circle from O to O is obtained by putting $\theta \rightarrow 2\pi$, so we have

$$\exp(2\pi ik)$$

in that case. In the second case $\theta \rightarrow -2\pi$ is used in going from O to O, again giving

$$\exp(2\pi ik) \ .$$

This can give constructive interference at O and that in turn can give rise to a large probability density $|\psi|^2$ there. This will be related to the presence or absence of resistance.

The presence of a uniform induction B orthogonal to the plane of the current will break the symmetry. By (3.17a) above, in the first case there is an additional phase factor

$$\exp(-i2\pi\Phi/\phi_e)$$

on going once around. Φ is the flux enclosed in the circuit. In the second case the complex conjugate factor applies.

At O the value of $|\psi|^2$ will be altered as B is varied. This can explain the negative magneto-resistance that is observed in some disordered conductors. The electrical resistivity will contain a term having the periodicity of the AB factor. Notice that the free movement backwards and forwards on a trajectory obeys time-reversal invariance T. The AB phase factors break T, and so their presence can be detected.

Similar phenomena can be observed in magneto-resistance experiments. For example a thin hollow metal cylinder encloses a narrow solenoid lying on the axis of the cylinder. The solenoid has magnetic flux Φ. It is found that the electrical resistivity (along the cylinder) will oscillate with the period $\phi_e/2$ as the flux is varied. For an account of resistivity oscillations see the article by A. G. Aronov and Yu. V. Sharvin [3.8]. This article also contains an account of relevant experiments.

3.5 The Basic Principle (General B_z Field)

Instead of emphasising Eqs. (3.18, 18a), we can return to the basic equation (2.2), and for an electron we have:

$$\boldsymbol{\Pi} = \boldsymbol{p} + e\boldsymbol{A} \ . \tag{3.19}$$

We assume that the momentum operators $\boldsymbol{\Pi}$ and $\boldsymbol{p}$ are associated with the wave functions $\psi_0(\boldsymbol{x})$ and $\psi_A(\boldsymbol{x})$ respectively. Equation (3.19) is not gauge invariant, and the simplest gauge-invariant functional of $\boldsymbol{A}$ is, by (2.16),

$$\oint_C \boldsymbol{A} \cdot \mathrm{d}\boldsymbol{s} \ , \tag{3.19a}$$

namely the integral of $\boldsymbol{A}$ around any closed curve C.

The next step is to restrict the curve C to be any plane closed orbit around which the electron wave packet moves under the influence of a static magnetic field B_z. [In Sect. 4.1 below it is shown that there are various closed paths C that are traversed solely under the influence of certain induction fields of the form $B_z(\rho)$]. The motion under the magnetic field is described by the Hamiltonian (2.4) and the wave function $\psi_A(\boldsymbol{x})$; the mean wavelength is λ_A.

We can choose a fairly narrow trench-like confining potential V_C, whose purpose is to keep the electron on (or near) the path C, without any induction field being present. In this case, if s and n are the tangent and normal to C, and n is the distance along n from C, the potential is of the form $V_C(n)$, and it is independent of s and z.

The speed of the electron is constant in the two cases. We conjecture therefore that the following is possible. The potential V_C is adjusted so that Schrödinger's equation arising from (2.4a) (in which there is no induction) gives a wave packet $\psi_0(\boldsymbol{x})$ having the same velocity and path C as the $\psi_A(\boldsymbol{x})$ solution describes. Let λ_0 be the mean wavelength for $\psi_0(\boldsymbol{x})$. The simplest gauge-invariant representation of (3.19) becomes

$$2\pi \oint_C \left(\frac{\mathrm{d}s}{\lambda_0} - \frac{\mathrm{d}s}{\lambda_A} \right) = (e/\hbar) \oint_C \boldsymbol{A} \cdot \mathrm{d}\boldsymbol{s} \ , \tag{3.19b}$$

where $\mathrm{d}\boldsymbol{s}$ is the element of arc. Equation (3.19b) expresses the phase difference between $\psi_0(\boldsymbol{x})$ and $\psi_A(\boldsymbol{x})$ in one circuit of C.

A more complicated derivation presumably gives the same result in the presence of a conservative potential (provided the path C does not approach any region where λ_A or λ_0 becomes large).

Thus, testing the AB phase factor precisely (as might be done in an interference experiment, (cf. Chap. 5)) will verify the basic relation (3.19). Only a gauge invariant functional of $A(x)$, as in (3.19 a or b), can be measured, and it is not easy to see how that measurement could imply any unusual property or significance, of the vector potential, for example in vacuo. In these considerations, we are only dealing with measurement and measured quantities.

3.6 The Hollow Field (Axially Symmetric B_z)

Equations (3.17)–(3.18a) show the interpretation of the wave-mechanical solution for an electron moving in a uniform field B. This is a special case because (a) the orbit is a circle, (b) the flux through the orbit is $\pi\rho_0^2 B$, whereas it could have other values. We can remove the last constraint by using the axially symmetric field

$$B_z = (\rho/\rho_T)^{n-1}B \quad , \tag{3.20}$$

where B and ρ_T are constants and $n \geq 0$ is an integer. On the circle $\rho = \rho_T$ the field is $B_z = B$. The details of the solution are given in Appendix A1. The vector potential A can be chosen to have only an azimuthal component $A_\theta(\rho)$, which only depends on ρ.

For $n > 1$ the field inside the circle $\rho = \rho_T$ is less than the value B, hence the name "hollow field". In such a case the AB effect will be reduced.

The solution for an electron wave packet moving on $\rho = \rho_T$ is of the form

$$\psi = \chi(\rho, t) \exp(iM'\theta) \quad , \qquad (M' = \text{integer}) \tag{3.20a}$$

and the extremum condition [of (A.4)] then yields

$$M'\hbar = \frac{n}{n+1}eB\rho_T^2 \quad . \tag{3.20b}$$

Equation (3.20b) differs from the condition in (3.4) for the uniform field case, $n = 1$. The electron moves in induction B; it will therefore have the cyclotron frequency ω_B of (3.5a), and the peripheral momentum is

$$P_\theta = eB\rho_T \quad , \tag{3.20c}$$

which is similar to the value in Sect. 3.1, for the case $n = 1$. The kinetic energy T_θ is given in (A.4b). It obeys

$$T_\theta = P_\theta^2/2m \quad ,$$

but (3.20b) shows that unless $n = 1$, T_θ *will not* equal $M'\hbar\omega_B$.

3.7 Equivalent Wave Functions

Consider the equivalent wave functions $\psi_A(x)$ and $\psi_0(x)$, with ψ_A given by (3.20a). Integrating once around the orbit $\rho = \rho_T$, we have by (A.2)

$$\oint_{(\rho_T)} A \cdot ds = \frac{2\pi B \rho_T^2}{n+1} \ . \tag{3.21}$$

Comparing with (3.18a) we see that the change in phase of $\psi_0(x)$ on going once around the orbit is

$$2\pi M' + \frac{e}{\hbar} \oint_{(\rho_T)} A \cdot ds = (eB\rho_T^2) \frac{2\pi}{\hbar} \ . \tag{3.21a}$$

Here (3.20a, b) and (3.21) have been used. The second term on the left side of (3.21a) is (minus) the AB phase. The term on the right of (3.21a) gives the phase of $\psi_0(x)$. It gives the angular momentum

$$\Lambda = eB\rho_T^2 \ . \tag{3.21b}$$

This is consistent with P_θ of (3.20c).

Equations (3.20b), (3.21) and (3.21a) thus show how, for different values of n, the total phase change $2\pi\Lambda/\hbar$ is made up of different proportions from the quantized term and the AB term. In each case the particle describes the same orbit $\rho = \rho_T$, in induction $B_z = B$. In the extreme case of $n \to \infty$, the AB term vanishes; while for $n = 0$, the quantized term vanishes.

3.8 The Betatron

Non-relativistic electrons, moving on a path of fixed radius ρ_T in induction B, can be accelerated merely by increasing B slowly and steadily, provided that the flux through the orbit $\Phi(\rho_T)$, always obeys

$$\Phi(\rho_T) = 2\pi \rho_T^2 B \ , \tag{3.22}$$

[see for example Sect. 2.4 of Ref. [3.9]]. The accelerator based on this principle is known as the betatron. Equation (3.21) shows that this condition is obeyed by the field in (3.20) for $n = 0$. (Clearly this is not a "hollow" field; rather the reverse.) Equation (3.22) does not specify the profile of the field inside $\rho = \rho_T$.

In Appendix A1 it can be seen that the quantum theory solution for $n = 0$ has special properties. If $M' \neq 0$, the expression (A.3a), with $A_\theta = B\rho_T$, has no minimum for finite ρ. Thus we must use $M' = 0$; consequently $\psi_A(x)$ does not vary with θ. Using a correspondence as in (3.18a), the wave function $\psi_0(\theta)$ is given by the AB phase factor $(\bar{U}_A(\theta))^{-1}$ alone. So we have

$$\psi_0(x) = \exp\{(ie/\hbar)\Phi(\rho_T)\theta/2\pi\}\chi(\rho, t) \tag{3.22a}$$

where $\chi(\rho, t)$ is the radial wave equation. Using (3.22), the angular momentum Λ is now given by

$$\Lambda_z \psi_0(\boldsymbol{x}) = \frac{\hbar \partial}{i \partial \theta} \psi_0(\boldsymbol{x}) = eB\rho_{\mathrm{T}}^2 \psi_0(\boldsymbol{x}) \ .$$

The eigenvalue Λ, which agrees with (3.21b), will increase slowly as B is slowly increased, provided (3.22) is preserved.

Equation (A.6b), or the analogous classical equation (B.5) of Appendix B, shows that there are no (radial) cyclotron oscillations in the case $n = 0$. The wave mechanics of the betatron is thus given by the AB phase factor alone.

For further discussion of induction fields in accelerators, see Sect. 8.11.2 below.

4. Classical Physics
and the Adiabatic Principle

In this chapter we shall use elementary classical arguments, plus the results of the classical mechanics adiabatic theorem, which is discussed in some detail in Sects. 8.2, 3. The purpose is to show properties that are related to the AB effect, as in Sect. 4.5, and to show the structure of the angular momentum Λ in the presence of magnetic induction, as in Sect. 4.6.

The signification of the several terms in the phase shift of de Broglie waves during an electron's motion in a magnetic field is made clear by looking at the classical solution for the motion. Moreover, apart from the solenoid case, the wave-mechanical solutions we have given above deal only with circular motion in given fields B_z. Such solutions miss an essential feature, namely the non-central nature of the Lorentz force. We saw that the original AB effect was easily understood by looking at the electromagnetic induction due to changing the flux in the solenoid. Similarly, the generalized AB effect is easily understood by looking at consequences of the non-central nature of the Lorentz force.

4.1 Motion in Axially Symmetric Induction

Consider the motion of a non-relativistic electron in an axially symmetric induction,

$$B_z = \frac{1}{\rho}\frac{d}{d\rho}(\rho A_\theta) \ , \tag{4.1}$$

where, in cylindrical coordinates (z, ρ, θ),

$$A = (0, 0, A_\theta) \ ; \qquad A_\theta = A_\theta(\rho) \ . \tag{4.1a}$$

Assume that the Lorentz force (2.1) is the only force acting on the electron. The electron's speed $|v|$ will then be constant.

4.1.1 The Nature of Closed Orbits

We are in some cases interested in closed electron paths, and it should be noted that other paths are possible besides the circles $\rho = \text{const}$. The curve in Fig. 4.1 is its own mirror image in the line $\theta = -\pi/2$, $\theta = \pi/2$. The derivative

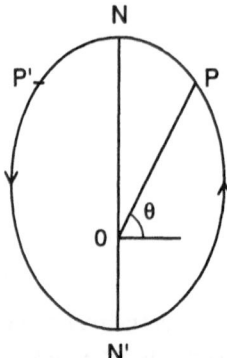

Fig. 4.1. Electron moving in the plane $z = 0$

$\rho' \equiv d\rho/d\theta$ obeys: $\rho' > 0$ in $-\pi/2 < \theta < \pi/2$, and $\rho' = 0$ at $\theta = \pm\pi/2$. Only one pair of points P, P' corresponds to each value of ρ in

$$ON' < \rho < ON \ .$$

For any given speed v, at each of these points we choose the induction

$$B_z(\rho) = \frac{mv}{eR} \ , \tag{4.2}$$

where R is the radius of curvature at P. (A suitably chosen and driven current density $j_\theta(\rho)$ can produce $B_z(\rho)$.)

A field defined in this way will guide the electron, of speed v, around the closed loop. Thus non-circular orbits do exist in axially symmetric fields of the form $B_z(\rho)$, so long as B_z is not uniform.

4.1.2 Integrals of Motion

The non-relativistic Newtonian equations of motion (in the plane $z = 0$) for a particle of charge $q = -e$ moving in an axial symmetric field $B_z(\rho)$ are:

$$\ddot{\rho} - \rho\dot{\theta}^2 = -\omega_B(\rho)\rho\dot{\theta} \ , \tag{4.3a}$$

$$2\dot{\rho}\dot{\theta} + \rho\ddot{\theta} = \omega_B(\rho)\dot{\rho} \ , \tag{4.3b}$$

where

$$\omega_B(\rho) = eB_z(\rho)/m \ .$$

Equation (4.3b) gives a first integral of motion:

$$m\rho^2\dot{\theta} - e\rho A_\theta = K \ . \tag{4.4}$$

K is a constant along the electron's path, and that path need not be a circle, or even a closed loop. The value of K is arbitrary. We can write

$$\Lambda - e\rho A_\theta = K \ , \tag{4.4a}$$

Λ being the (ordinary) angular momentum about Oz.

Equation (4.4a) is called the law of conservation of *canonical angular momentum*. The relation shows very neatly how Λ is forced to vary by the non-central nature of the Lorentz force. If the path is not a circle $\rho = \text{const}$, then, except for the case of the solenoid, ρA_θ will vary, and so must Λ.

Notice that if $\boldsymbol{A}$ has the form of (4.1a), then $A_\theta(\rho)$ can be expressed as a gauge invariant quantity [because it equals $(2\pi\rho)^{-1}$ times a flux]. So (4.4a) and therefore K itself, are gauge invariant.

Since (4.4a) is closely related to (3.21a) for the composition of the wave mechanics phase factor, Eq. (4.4a) is also important for the generalized AB effect.

A second integral of motion that follows from (4.3a, b) is

$$m\left(\dot{\rho}^2 + \rho^2\dot{\theta}^2\right) = 2T \tag{4.4b}$$

where T is a positive constant for the chosen path.

4.2 Action Variables

A proper account of the angle and action variables for cyclic or periodic motions is given in Sect. 8.2. Here we give a simple discussion.

The Lagrangian for this problem is

$$L = \tfrac{1}{2}mv^2 - e\boldsymbol{v} \cdot \boldsymbol{A}$$
$$= \tfrac{1}{2}m\left(\dot{\rho}^2 + \rho^2\dot{\theta}^2\right) - e\rho\dot{\theta}A_\theta \quad . \tag{4.5}$$

The action of a conservative system having generalised coordinates q_r ($r = 1, 2, \ldots N$) is

$$A = \frac{1}{2\pi} \oint \sum_{r=1}^{N} \dot{q}_r \frac{\partial L}{\partial \dot{q}_r} \, dt \quad , \tag{4.6}$$

where the integration is over one period of the system, or once around a possible loop. (These two cases are called libration or rotation, respectively.) The symbol A is used for the general expression (4.6) and the parts of A relating to single coordinates will be called *action variables* and labelled J etc. Thus in our example

$$\frac{\partial L}{\partial \dot{\theta}} = m\rho^2\dot{\theta} - e\rho A_\theta \quad ,$$

so we define an action variable J_θ to be

$$J_\theta = \frac{1}{2\pi} \oint \dot{\theta}\frac{\partial L}{\partial \dot{\theta}} \, dt = \frac{1}{2\pi} \oint \left(m\rho^2\dot{\theta} - e\rho A_\theta\right) d\theta \tag{4.7}$$

as is also used in (8.51). The integration is around a closed path, and in general that will not be a circle. Note that (4.5) is a special case, in that L

does not explicitly contain θ. Therefore, by Lagrange's equation, $\partial L/\partial \dot{\theta}$ does not vary with time, and (4.4) or (4.4a) is a consequence.

Hence for a closed orbit

$$J_\theta = K$$

by (4.7) [θ is an angle variable in the sense of (8.11)].

Another action variable is

$$J_\rho = (m/2\pi) \oint \dot{\rho}^2 \, dt = (m/2\pi) \oint \dot{\rho} \, d\rho \ . \tag{4.7a}$$

Here the integration is over a period of the motion; cf. (8.8).

Clearly $J_\rho \geq 0$, and $J_\rho = 0$ gives a circle $\rho = $ const. Thus if

$$\rho(t) = \rho_T + D \cos \omega t \ ,$$

where D is a constant and ω is any frequency,

$$J_\rho = \frac{m}{2\pi} (D\omega)^2 \int_0^\tau \sin^2 \omega t \, dt \ ,$$

where $\tau = 2\pi/\omega$ is the period of this vibration. Hence

$$J_\rho = \tfrac{1}{2} m\omega D^2 \ . \tag{4.7b}$$

Examples

(i) For motion on the circle $\rho = \rho'$ in the field of (3.20), on using (4.7), (4.7a) and (A.2) we find,

$$J_\rho = 0 \ , \qquad J_\theta = \frac{n}{n+1} eB(\rho')\rho'^2 = \frac{n}{n+1} \Lambda(\rho') \ . \tag{4.8}$$

Here $B(\rho')$ is the induction on $\rho = \rho'$, and $\Lambda(\rho')$ is the angular momentum.

(ii) *Around the Solenoid.* The same derivation of a constant of motion K holds under the influence of a conservative potential V, *provided V does not depend on θ*. By (2.3) and (4.4a), motion on a closed path around the solenoid (Fig. 2.3) under a guiding potential $V(\rho)$ gives the relation

$$K = \Lambda - \frac{e}{2\pi} \Phi \ , \tag{4.8a}$$

where Φ is the flux in the solenoid, and Λ is the angular momentum. (The orbit does not have to be a circle.)

Equation (4.8a), written in the form

$$\Lambda = K + \frac{e}{2\pi} \Phi \ , \tag{4.8b}$$

can be regarded as the classical form of the AB relation (cf. Sect. 4.5 below for further comments on this point).

Case Where $V(x)$ *Depends on* θ. In the case where $V(\rho,\theta)$ varies with θ, (4.4) will not be an integral of motion, and the constant K will not exist. However the action variable J_θ defined by (4.7) will exist for closed orbits. The adiabatic theorem (cf. Sect. 4.4 below, and in more detail in Sect. 8.3) then states that J_θ does not change when the flux Φ is varied sufficiently slowly.

Thus for any closed orbit around the solenoid, Eq. (4.8b) is replaced by the relation

$$\frac{1}{2\pi} \oint m\rho^2 \dot\theta \, d\theta = J_\theta + \frac{e}{2\pi}\Phi \ , \tag{4.8c}$$

where J_θ is constant. This result is also a more general form of the result in Sect. 2.7.

Comment on Equation (4.8b). For motion around the solenoid, in the case where the confining potential V is independent of θ, the argument we have just given shows that in (4.8b), K is an adiabatic constant. Thus K does not vary when the flux Φ is changed sufficiently slowly.

4.3 Angle–Action Equations of Motion

A conservative periodic dynamical system such as we are considering, can be described in terms of canonical conjugate pairs of variables:

$$(J_\theta, \phi_\theta) \ , \qquad (J_\rho, \phi_\rho) \ .$$

ϕ_θ, ϕ_ρ are angle variables, but in general they are not geometric angles. The solutions of (4.3a, b) that give closed orbits are described by constant values of J_θ and J_ρ. The Hamiltionian H can be expressed as a function of the action variables: $H = H(J_\theta, J_\rho)$ and H does not depend on ϕ_θ or ϕ_ρ.

The angle variables obey the equation of motion,

$$\dot\phi_\theta = \frac{\partial H}{\partial J_\theta} \ , \qquad \dot\phi_\rho = \frac{\partial H}{\partial J_\rho} \ . \tag{4.9}$$

Thus the position of the electron on the orbit is related to ϕ_e and ϕ_ρ, and both of them advance uniformly with time, with the rates shown in (4.9).

For motion under the axially symmetric induction field $B_z(\rho)$ of (3.20), it is shown in Appendix B how H is expressed in terms of J_θ and J_ρ alone. Choose a number $n > 0$. Then corresponding to a positive number K there is a radius ρ', such that motion on the circle $\rho = \rho'$ gives K, by (4.4a). Since this is a closed orbit, by (4.7), $K = J_\theta$. Thus, as in (B.3), ρ' is a function of J_θ, and vice versa.

For $J_\rho = 0$ the motion is on the circle $\rho = \rho'$. For $J_\rho > 0$, the value of ρ oscillates between ρ_1 and ρ_2. According to the approximation used in Appendix B, (B.6a), we have

$$\rho_2 - \rho' = \rho' - \rho_1$$

$$= n^{-1/4} \left(\frac{2J_\rho}{eB(\rho')} \right)^{1/2} . \tag{4.10}$$

$B(\rho')$ being the field on $\rho = \rho'$. Oval-shaped orbits are specified (partially) by this relation.

From (4.9), motion on the circle $\rho = \rho_T$ gives $\dot{\phi}_\theta = \omega_B$ [as in (3.5a)], and for small radial oscillations about $\rho = \rho_T$ we find

$$\dot{\phi}_\rho = n^{1/2}\omega_B$$

in agreement with (A.6b) of Appendix AI, and (B.5) of Appendix B.

4.4 The Adiabatic Principle

It is a striking property of a conservative dynamical system that if any parameter of the system, such as B in (3.20), is altered slowly then the energy of the system, and other properties of the system, will change slowly, but the action variable, or variables, will stay approximately constant. A common example is a steel ball swinging under gravity on an inelastic string of length l. On reducing l slowly the frequency ω will increse as $l^{-1/2}$. The invariance of the action will ensure that the angular amplitude of swing Θ will vary as $l^{-3/4}$. In this way the adiabatic principle relates one dynamical system to another dynamical system having different values of the parameters.

This stationary property of the action variable under adiabatic change is proved in Sect. 8.3.

The adiabatic principle (as shown in Sect. 8.3) does not give a simple prescription for how slow any variation of the parameters has to be in order to be adiabatic.

A degeneracy in several frequencies, occurring as a parameter $a(t)$ varies, can invalidate the result; as can the situation in which a frequency ω goes to zero. Several proofs of the adiabatic theorem, stating various other exceptions and conditions, appear in standard texts, e. g. E. Fues [4.1], V. I. Arnold [4.2].

As an example, consider the motion of an electron (in the plane $z = const.$) on a circle of radius $\rho = \rho'$ under the induction in (3.20). B is the induction on the standard circle $\rho = \rho_T$. Keep ρ_T fixed and vary B slowly. For $n > 0$, (4.8) shows that ρ' varies so that

$$B(\rho')^{n+1} = \text{const.}$$

It follows that the angular velocity $\omega(\rho')$ and the kinetic energy T_θ both vary as

$$(B)^{\frac{2}{n+1}} .$$

Uniform induction is given by $n = 1$.

The "magnetic moment" of an orbit is $ev_\perp^2/2\omega(\rho')$, where $v_\perp$ is the velocity component orthogonal to B. This equals T_θ/B, and for $n = 1$ it is an adiabatical invariant.

Equation (4.8) gives no information for $n = 0$; that is the betatron case discussed in Sect. 3.8.

4.5 Relation to the AB Effect

Consider an electron moving around the solenoid in Fig. 2.3. Equation (4.8b) shows how the phase factor

$$\exp(i\Lambda\theta/\hbar) \tag{4.11}$$

of the equivalent "field-free" wave function $\psi_0(x)$ will behave when the flux Φ is slowly changed from zero to its final value. As was mentioned in Sect. 4.2, J_θ will not alter. (And here J_θ equals K, since the orbits are closed, and the confining potential V is independent of θ.) Therefore the dependence of this phase factor on Φ is

$$\exp(ie\Phi\theta/h) = \exp(i\Phi\theta/\phi_e) \quad . \tag{4.11a}$$

This is the original AB effect.

Consider the generalized AB effect in the axially symmetric field $B_z(\rho)$ of (3.20). We look at a wave packet moving on the standard circle $\rho = \rho_T$. The phase relation between $\psi_0(x)$ and $\psi_A(x)$ for these wave packets is given by (3.21) and (3.21a). With the notation of (3.21b) the relation is

$$\Lambda = \frac{e}{2\pi} \oint_{(\rho_T)} A \cdot ds + M'\hbar \quad . \tag{4.12}$$

Now (4.12) is the same form as the classical relation (4.4a) for the action variable J_θ $(= K)$. The two relations are identical on putting $K = M'\hbar$, and on letting ρ_T be one of the allowed quantized values given in (3.20b). (This last step is equivalent to a Bohr–Sommerfeld quantization.)

The details of this calculation are given in Sect. 8.10, in particular the argument leading to (8.55).

4.6 The Variation of Λ

The interplay between angular momentum Λ and the quantity K is best seen in non-circular motion [in an axially symmetric field $B_z(\rho)$] where K is the constant in the integral of motion (4.4a). The Lorentz force, (2.1), deflects the moving electron, and thereby alters Λ.

This phenomenon is easily seen from the simple scattering shown in Fig. 4.2. An electron coming from far away is deflected by an axially symmetric field $B_z(\rho)$. The induction falls off to zero smoothly outside the batched

region around O. We choose $B_z(\rho) \geq 0$, so the Lorentz force always deflects
the electron to the left. The angular momentum at the points

$$-\infty, \quad U, \quad W, \quad V, \quad +\infty \ ,$$

is respectively

$$mvb, \quad 0, \quad -mv\rho_m, \quad 0, \quad mvb \ ,$$

where v is the velocity and b, ρ_m are the pedal distances shown in Fig. 4.2.
The figure shows the case $b > 0$. The tangents to the path at U and V pass
through O.

It is easy to deduce from (4.2) that the angle γ between the velocity and
a fixed direction obeys

$$\frac{d\gamma}{ds} = \frac{\omega(\rho)}{v} \ ,$$

where v is the (constant) speed, ds is an element of arc, and $\omega(\rho)$ is defined
in (B.2b). The total angle of rotation over a length of arc s, described in time
T is

$$\gamma(s) - \gamma(0) = v^{-1} \int_0^s \omega(\rho') \, ds' = \int_0^T \omega(\rho') \, dt' \ .$$

By choosing v sufficiently large we can arrange that the angle Ψ in Fig. 4.2
is less than π. Otherwise the diagram, and the phenomena, become more
complicated than we wish to discuss here.

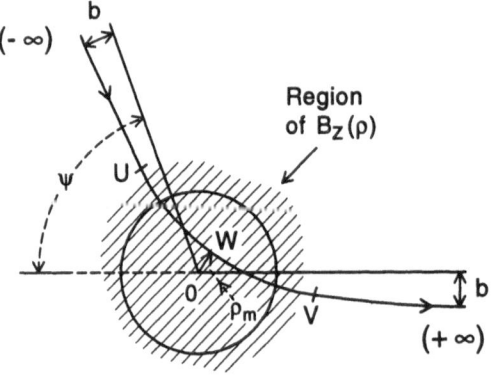

Fig. 4.2. Passage of an electron through a region of field
$B_z(\rho)$. Initial and final angular
momenta are mvb. Λ vanishes
at U and V

In this axially symmetric field B_z, with no potential $V(x)$ present the
integral of motion in (4.4a) can be used. Therefore Λ is a function of ρ and
it obeys

$$\Lambda(\rho) - \frac{e}{2\pi} \oint_{(\rho)} \boldsymbol{A} \cdot d\boldsymbol{s} = \Lambda_\infty - \frac{e}{2\pi}\Phi(\text{tot.}) \equiv K \tag{4.13}$$

where K is a constant. The second term on the left is written as an integral around the circle of radius ρ (centered on O) in order to make the gauge invariance explicit. We have $\Lambda_\infty = mvB$, and Φ(tot.) is the flux through a circle so large that on it B_z vanishes. Since $B_z \geq 0$, when the electron moves in form $-\infty$, the flux decreases from Φ(tot.). Thus Λ decreases to zero at U, and goes through negative values until W is reached, with the minimum value $\Lambda(\rho_{\mathrm{m}}) = -mv\rho_{\mathrm{m}}$.

In (4.13) K is the first constant of motion, and here it is *not* an action variable since the motion is neither cyclic nor periodic. However the form of (4.13) shows that an expression analogous to the AB term is an integral part of any simple description of the asymmetrical motion of a charged particle in an induction field.

The fact that Λ, the component of angular momentum about Oz, is not constant for motion in the induction field B_z, causes K, the integral of motion constant, to be important. The difference $(\Lambda - K)$ is similar to what we called the generalised AB effect, though in the present case the motion is aperiodic.

5. Problems with, and Criticisms of, Experiments and the AB Effect Itself

We discuss the structure of various experiments, and also experimental topics that are closely related to theory. Other topics are included that have a bearing on some of the numerous discussions on the existence of the AB effect and its nature. There is no attempt here to give a history of all the objections (to the AB effect itself and to various experiments), alternative explanations, corrections of objections, etc., that are in the literature. Good surveys of these can be found in [5.1] and [5.2] and elsewhere.

Biprism electron interference experiments, and the limitations on such measurements, are discussed, together with the problem of distinguishing the fringe shift from the profile shift. The use of tapered magnetic whiskers is described, together with Pryce's leakage induction field, and the important arguments to which it led. Finally, the relation between the AB effect and the superconductor interference experiments, and the quantization of flux in superconductors is examined.

5.1 The Standard Biprism Experiment

Figure 5.1 shows schematically the standard electron interference experiment for the AB effect. Electrons from a small coherent source S pass on routes 1 and 2 on each side of the cylindrical solenoid, or cylindrical magnetic fibre. The axis of the solenoid, or fibre, is parallel to O_z, which is vertically up out of the paper. The electrons move in, or close to, the plane Oxy, and they are electrostatically deflected near D_1 and D_2, so that the two wavelets meet and interfere at Q on Q_{yz}, which is the observation plane. Details of such experiments are given in Refs. [5.3]–[5.8] and a review in [5.9].

The equivalent electron optics description with rectilinear propagation and mutually coherent sources S_1, S_2 is shown in Fig. 5.2. When $\boldsymbol{B} = 0$ in the solenoid, the intensity observed at Q is

$$\tfrac{1}{2}|\psi_1 + \psi_2|^2$$

where ψ_1, ψ_2 are the wave packets (of mean wavelength λ) passing via D_1, D_2 respectively. If ψ_1 and ψ_2 equal $\exp(\mathrm{i}\alpha)$ and $\exp(\mathrm{i}\beta)$ at point Q (with α, β real), the maxima of the interference pattern on Oyz occur where

$$\alpha - \beta = \pm 2n\pi \qquad (n = 0, 1, 2, \ldots) \ . \tag{5.1}$$

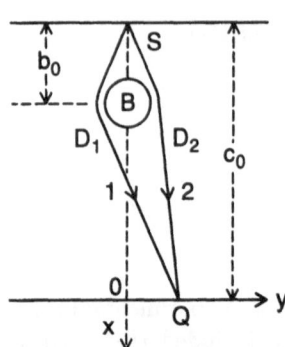

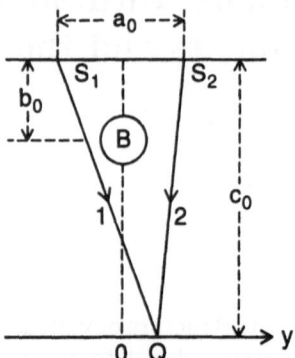

Fig. 5.1. Biprism electron interferences (angles exaggerated)

Fig. 5.2. Equivalent rectilinear paths (angels exaggerated)

If there is flux Φ in the solenoid, the extra phase factor $U(2\pi)$, of (2.7a) and (2.9) will be present and the maxima occur where

$$\alpha - \beta - 2\pi\Phi/\psi_e = \pm 2n\pi \qquad (n = 0, 1, 2, \ldots) \ . \qquad (5.1a)$$

It can be seen from Fig. 5.2, that for $\Phi > 0$ the fringe pattern is shifted to the right. [This agrees with the sign of the current density j_θ in (2.15b).] The shift is

$$y = \Delta_f \frac{\Phi}{\phi_e} \qquad (5.2)$$

where Δ_f is the fringe separation. With the parameters in Fig. 5.2

$$\Delta_f = c_0\lambda/a_0 \ . \qquad (5.2a)$$

A survey of typical biprism experiments is given in Ref. [5.9]. Typical values of the parameters are:

$\lambda = 9 \times 10^{-3}$ nm (20 keV electron), $\quad a_0 = 10\,\mu$m

$b_0 = 4$ cm, $\quad c_0 = 30$ cm, $\quad \Delta_f = 340$ nm

solenoid radius = a few μm .

It may be useful to notice that for a solenoid radius $r_f = 2\,\mu$m, a flux equal to ϕ_e of (2.8) is given by an induction of 3 Gauss in the solenoid.

5.2 Types of Measurement

The measurements which have been made are not absolute. The experimenters do not work out precisely the geometry of the biprism apparatus of Fig. 5.1, and then calculate that the point O (see Fig. 5.1) is an equal electron-optical distance from the source S via rays 1 and 2. That would not be practical, for various reasons. A result of this is that only a shift of fringes Q due to a difference in flux in the solenoid in two experiments can be determined.

We shall call the value of the flux in the solenoid the setting of the apparatus. If the apparatus is switched on at one setting Φ_1, and an interference pattern is formed, and then the same is done at another setting Φ_2, we cannot use the shift of the fringes (y) in (5.2) to give $\Phi_2 - \Phi_1$. This is because individual fringes are identical. Only the shift

$$\Phi' = \Phi_2 - \Phi_1 (\mathrm{mod}\, \phi_e)$$

of (2.17b) can be determined by such a measurement. An experiment like this, designed to compare the fringes at two settings, requires at least a moderately large number of electrons, as we discuss in Sect. 6.5 below.

We can measure flux differences very much larger than ϕ_e by measuring the fringe positions for numerous settings Φ_j that obey:

$$\Phi_j < \Phi_{j+1} \quad , \qquad |\Phi_{j+1} - \Phi_j| < \tfrac{1}{4}\phi_e \quad , \qquad (\text{all } j) \ .$$

It should be possible in this fashion to measure a shift y that is much larger than Δ_f.

In view of the analysis in Sects. 2.7, 2.9 above, this latter experiment can be regarded as demonstrating electromagnetic induction, rather than the AB effect. The experiment would require using very large numbers of electrons. It is an example of the use of a very large number of quanta to overcome quantum theory limitations and thereby approximate to a macroscopic result. See Sects. 6.3–6.5 below for other examples of this technique.

5.3 Must the Electron be Present?

This question relates to one of the philosophical problems that have arisen from the AB effect. In the case of the solenoid, if the change of the flux Φ is made while no electron is present, how does an electron that arrives after the change, perceive the change? In classical theory an electron that was not present while the change of flux occurred will not be subject to any effect, because before and after the change there is no field outside the solenoid.

In quantum theory there are two separate concepts: the state ψ, and its occupation number n, which may be 0 or 1. Changing the flux Φ changes the state ψ, whether n is 0 or 1. The change in ψ is given by an alteration in the relation between $\psi_A(x)$ and $\psi_0(x)$ that was shown in Sect. 2.10; changing Φ can thus be seen as a change in the boundary condition that the wave function must obey on going once around the surface of the solenoid.

The situation is somewhat similar to the polarization of the vacuum around an atomic nucleus of charge $+Ze$. Changing Z will change the polarization whether or not any electron is present. The difference between the two cases is that it may be easier to envisage the electric field arising from the vacuum polarization, than it is to remember the difference between wave functions $\psi_A(x)$ and $\psi_0(x)$. The use of the generalized AB effect in Chap. 3, and the classical relation between Λ and the action variable K in Chap. 4, should make it easier to understand the roles of $\psi_A(x)$ and $\psi_0(x)$. It may not be easy to envisage an alteration in the change of phase on going once around the solenoid; however we emphasise that this is a well-defined concept.

For an early discussion of this topic see the appendix to Ref. [5.10].

It is relevant to mention here that some physicists have raised the question of what happens to an AB interference when the flux that causes the interference is suddenly altered, or turned off [5.11]–[5.13]. Bearing in mind (i) the effect of Faraday's induction, discussed in Sect. 2.7; and (ii) the fact that electromagnetic signals cannot propagate faster than c, there should be no paradox or other problem in such phenomena. Causality is not violated.

5.4 Fringe Shift ad Profile Shift

In the notation of Sect. 5.1 the intensity at point Q in the apparatus of Fig. 5.1 is

$$\tfrac{1}{2}|\psi_1(y)|^2 + \tfrac{1}{2}|\psi_2(y)|^2 + |\psi_1(y)|\,|\psi_2(y)|\cos(\alpha - \beta - 2\pi\Phi/\phi_e) \ . \qquad (5.2b)$$

The profile of the fringe pattern is the area bounded by

$$\pm|\psi_1(y)|\,|\psi_2(y)| \ .$$

$|\psi_1(y)|\,|\psi_2(y)|$ can only be non-zero provided that the rays 1 and 2 can pass the solenoid and still reach Q, and provided also that the coherence length l_C of the electron wave packet coming from S is sufficiently large that the two rays can overlap at Q.

Equations (2.6) and (5.2b) show that the AB phase can shift the fringes, but it does not, by itself, move the profile. The paper [5.14] gets this last aspect wrong. The profile is an intensity correlation effect whereas the AB effect is only a phase effect.

5.4.1 Kirchhoff-Type Calculations

The diffraction, or scattering, of two coherent electron wave packets from sources S_1 and S_2 past a circular cylinder, as in Fig. 5.3, has been calculated by Kirchhoff's technique. The Fresnel edge of the ray from S_1 lies to the *right* of the Fresnel edge of the ray from S_2, and the interference pattern appears between the two edges on the observation plane. For details of these calculations see pp. 392–397 of Ref. [5.9].

Next a uniform induction field B_z' is applied over a rectangle whose width in the direction O_x is a'. This field is close to the cylinder, and the two rays 1, 2 pass through it, as shown in Fig. 5.3. The Kirchhoff calculation now shows that the interference pattern, *both profile and fringes*, is shifted to the right by

$$y(B_z') = B_z'a'a_1\Delta_f/\phi_e \ . \qquad (5.3)$$

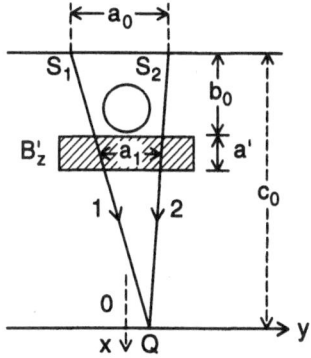

Fig. 5.3. Magnetic field between the paths of rays

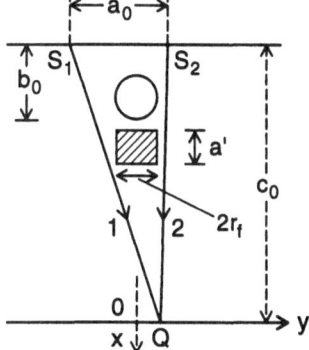

Fig. 5.4. The flux which was removed from Fig. 5.3

Δ_f is the fringe separation and a_1 is the distance between the two rays where they pass through the field B'_z. The shift of the fringes agrees with (5.2) for $\Phi = B'_z a' a_1$, which is the flux enclosed by the two rays.

There is a subtle point here. The cylinder in Fig. 5.3 has diameter $2r_f < a_1$. The position of the flux shown in Fig. 5.4 is in the "shadow" of the cylinder. The dimensions given just after (5.2a) should be remembered. Thus $b_0 \simeq$ 4 cm, $a_0 \simeq 10\,\mu$m, $\lambda \simeq 10^{-11}$ m, and $2r_f$ is (say) 5 μm. Hence the shadow of the cylinder (from the sources) is well defined.

If the field B'_z in Fig. 5.3 is removed from the volume in the shadow of the cylinder, i.e. from a rectangle of size $a' \times 2r_f$, then a further Kirchhoff calculation [5.9] shows that the profile of the interference pattern is shifted by $y(B'_z)$ of (5.3) as before, but the shift of the fringes is now less, namely

$$y' = B'_z a'(a_1 - 2r_f)\Delta_f/\phi_e \quad . \tag{5.3a}$$

Some authors suggest that the effect of the "hidden flux" in Fig. 5.3 is rather strange. The "hidden flux" is shown in Fig. 5.4, and it cannot move the profile since the rays do not pass through the induction. The flux $\Phi' = B'_z 2r_f a'$, between paths 1 and 2, simply increases the phase shift from the value in (5.3a) to that in (5.3).

5.5 Measuring Fringe Shifts Versus Profile Shifts

Various experiments have been carried out to demonstrate this phenomena (see [5.4] and [5.9] for a survey). We should mention a difficulty that may arise. Thermionic electron sources can have sufficient coherence to produce interference patterns of a few hundred fringes in apparatus of the biprism type. We shall consider 10–15 or more fringes in the pattern, and will assume that the pattern is limited by coherence only.

The objective is to compare two patterns having the same profile, and determine the shift of the fringes in one case relative to the other. The rule of the game is that we do not see the profile, we can only infer it from the fringes.

If the pattern is very broad we cannot detect a shift relative to the profile. A monochromatic wave has no structure.

Suppose the oscillation of the intensity above a flat background is

$$A(y) = \exp(-\sigma^2 y^2/2) \sin(2\pi y/\lambda_f) \tag{5.4}$$

where λ_f is the wavelength of the pattern and σ is a constant. A shift δ in the fringes gives

$$A'(y) = \exp(-\sigma^2 y^2/2) \sin(2\pi(y+\delta)/\lambda_f)$$

whereas a shift δ in profile and fringes gives

$$A''(y) = \exp(-\sigma^2(y+\delta)^2/2) \sin(2\pi(y+\delta)/\lambda_f) \ .$$

If $\exp(-\sigma^2 y^2/2)$ falls to half height in $N/2$ fringes then

$$\sigma\lambda_f \simeq \frac{2.4}{N} \ .$$

The amplitudes $A'(y)$ and $A''(y)$ differ, approximately, by the factor

$$\exp(-\sigma^2 y\delta) \ . \tag{5.4a}$$

At half height

$$\sigma^2 y\delta = 2.8 \frac{\delta}{N\lambda_f} \ .$$

The maximum shift to consider is $\delta = \lambda_f/2$, and in that case (5.4a) gives

$$\exp(-1.4/N) \ . \tag{5.4b}$$

N is approximately the number of fringes, and even for $N = 10$ it would be difficult to distinguish $A'(y)$ from $A''(y)$ by simple methods.

It should also be noted that in order to detect a fringe–profile shift effect in a convincing fashion, it is necessary to be sure that there is no asymmetry in the apparatus; that neither ray is close to the Fresnel edge; and so on.

It has also been suggested [5.9] that wave packets with very short coherence lengths l_C could be used, so that N is small, and the difference between a pattern with one strong central fringe and a pattern with two strong central fringes, would indicate a relative shift of $\lambda_f/2$.

5.6 The Chambers–Pryce Effect

In the first successful experiment Chambers [5.5] produced evidence of the AB effect in a biprism experiment like that shown in Fig. 5.1. However he used thin iron whiskers that were single magnetic domains of about 0.5 mm length and about 1 µm diameter, in place of the ideal cylindrical solenoid. The whiskers were slightly tapered, with a taper slope of 10^{-3} rad. This tapering showed up the AB effect in an elegant way.

The experiment with the whisker produced interference fringes as shown in Fig. 5.5. The first idea was that the fringes were tilted because the flux $\Phi(z)$ at height z in the whisker decreases, or increases, as z varies. The quantity

$$z_f = \phi_e \left(\frac{d\Phi}{dz} \right)^{-1} \tag{5.5}$$

is the length along the whisker in which Φ alters by the unit of flux ϕ_e. This distance z_f is closely related to the separation along the z-axis of the fringes shown in Fig. 5.5.

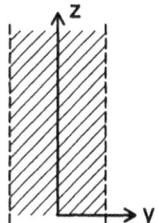

Fig. 5.5. Tilted fringes

The electron beams from the source S in Fig. 5.1 are in the plane $z = 0$, or close do it. If the pattern in Fig. 5.5 is about $30\Delta_f$ in the z-direction, the data after (5.2a) show that the deviation of rays from the $z = 0$ plane need not exceed 2×10^{-5} rad. Assuming the source behaves as a pinhole, the separation of the observed fringes along the z-axis is

$$\frac{c_0}{b_0} z_f \quad .$$

The slope of the fringes from the z-axis will be

$$\frac{b_0}{c_0} \frac{\Delta_f}{z_f} \quad . \tag{5.5a}$$

Typical values are $z_f = 1\,\mu m$, and the slope in (5.5a) may be 1/5 to 1/20 in practical cases. We assume that $|d\Phi/dz|$ is small enough that this slope is appreciably less than unity.

The difficulty about this explanation of the tilted fringes, is that, as Maurice Pryce pointed out [5.5], there has to be an induction field outside the tapered whisker. To a good approximation the leakage is radial, and by Gauss's theorem

$$B_\rho = -\frac{1}{2\pi\rho}\frac{d\Phi}{dz} \quad . \tag{5.6}$$

The radial induction B_ρ does not alter the flux through any circuit lying in a plane $z = $ const. However its presence prevents the derivation of the AB phase which is quoted in Sect. 2.3. Pryce therefore treated the effect of B_ρ as a perturbation, and he argued in a semi-classical fashion.

The Lorentz force of (2.1) due to the induction B_ρ, acting on an electron moving approximately in a plane $z = $ const., gives it a momentum δp_z in the z-direction. It is clear that this force only acts near the whisker, in the case of the experiment in Fig. 5.1. When the electron, on path 1 (Fig. 5.1), moves through an angle $\delta\theta$ about the axis of the whisker, the acquired impulse is

$$\delta p_z = -\frac{e}{2\pi}\frac{d\Phi}{dz}\delta\theta \quad . \tag{5.6a}$$

The effect on path 2 is numerically the reverse, and in each case we have to use $\delta\theta = \pi$.

At this point Pryce goes over to wave mechanics. That does not specify the phase of the wave function, and without any arbitrariness we can take the form in (5.2b) for the behaviour in the plane $z = 0$ [with $\Phi = \Phi(z = 0)$]. On a plane $x = -f$ which lies just beyond the whisker (and on the same side as O) in Fig. 5.1, the acquired impulses (5.6a) cause the wave packet on path 1 to acquire an extra phase

$$-\frac{e}{2}\frac{d\Phi}{dz}\frac{z}{\hbar} \tag{5.6b}$$

while the wave packet on path 2 gets the opposite extra phase. Thus on $x = -f$ when we move away from the plane $z = 0$, the phase difference $(\alpha - \beta)$ is altered by

$$-2\pi\frac{d\Phi}{dz}\frac{z}{\phi_e} = -2\pi\frac{z}{z_f} \quad . \tag{5.7}$$

On the observing plane $x = 0$ there is a trivial complication arising from the amplification factor c_0/b_0 mentioned above. It is simplest to say that on $x = 0$ the argument of the cosine in (5.2b) can be written as

$$\alpha - \beta - 2\pi\frac{\Phi[z']}{\phi_e} \quad , \tag{5.7a}$$

where

$$\Phi[z'] = \Phi(z = b_0 z'/c_0) \quad .$$

Equations (5.7) and (5.7a) will give the value in (5.5a) for the slope of the fringes in Fig. 5.5. Also, provided that $d\Phi/dz$ alters little over a distance of several times z_f, the extra phase in (5.7) will, in (5.2a), alter the phase

$$-2\pi\Phi(z = 0)/\phi_e \quad \text{to} \quad -2\pi\Phi(z)/\phi_e \quad .$$

5.6.1 Uncertainty and a Limitation

There is a restriction on this argument by Pryce. The wave function arriving at the whisker should cover a band δz in the z-direction that is sufficiently large that diffraction does not spoil the deflection estimated in (5.6a). We use the symbol Δ_{70} to denote a spread such that there is a 70% chance (in the Gaussian case) of the event lying within it [cf. (6.3) below]. The standard uncertainty relation becomes

$$\Delta_{70}z \times \Delta_{70}p_z \gtrsim h/\pi \ , \tag{5.8}$$

with the equality holding for a Gaussian minimum uncertainty profile.

We arbitrarily choose the condition:

$$\Delta_{70}p_z < \frac{1}{3}(\delta p_z)_{2\pi} \ , \qquad \text{with} \quad (\delta p_z)_{2\pi} = e\frac{d\Phi}{dz} \ .$$

This should ensure the validity of Pryce's argument. Equations (5.5) and (5.8) now require,

$$\Delta_{70}z > |z_f| \ . \tag{5.8a}$$

The width of the wave front at the whisker should therefore exceed $|z_f|$ if (5.7a) is to be valid.

This result can be understood in another fashion. Scanning along a line $y = \text{const.}$ in Fig. 5.5, fringes appear at separation $(c_0/b_0)|z_f|$, and that is equivalent to separation $|z_f|$ on $x = -f$. The coherent wave front should cover at least one such unit, in order that the fringes remain stable with respect to each other.

This consideration leading to (5.8a), appears to invalidate an argument, or objection, that Aharonov and Bohm [5.10] raise against Pryce's derivation of the tilted fringe result.

5.7 Potentials and Fields

It was pointed out in Chamber's paper [5.5] that, provided there is a position z_0 on the whisker where $\Phi(z_0) = 0$, then by successive use of (5.7), the whole AB phase at any point z on the whisker,

$$-2\pi\Phi(z)/\phi_e \ ,$$

can be built up by a set of applications of Pryce's effect. In this fashion the AB phase is produced by the leakage induction B_ρ outside the whisker. This was contrary to the opinion of Aharonov and Bohm [5.10], [5.15] that the AB phase was (solely) an effect of the vector potential. Various other authors have the same opinion. For discussion on this and related topics see the articles in [5.16], and the surveys in [5.9], [5.17].

The material presented in the present article will lead to the following scenario. The AB phase arising from the cylindrical solenoid discussed in

Chap. 2 is indeed simply derived from the potential A_Φ of (2.3); there appears to be no other simple derivation. However, if the flux Φ in the solenoid is slowly altered in time, electromagnetic induction, as in Sect. 2.7, gives the new AB phase. Similarly if the solenoid is no longer cylindrical but is finely tapered then Pryce's effect gives the AB phase. This analysis is further exploited in Sects. 5.8, 5.9 below.

One further aspect should be mentioned. Discussions on the origin of the AB effect led Aharonov and Bohm [5.10], [5.15] to the proposition that the potentials (like A) are more fundamental than the fields (like B). Many would agree with this if it were reworded to give something like: "the only simple and powerful method of writing Schrödinger's equation for the motion of an electron in an electromagnetic field, or of writing the equations of quantum electrodynamics, is by the use of the potentials $A_\mu(x)$, rather than the fields".

However, in practical terms, the *basic* reason that A occurs in the AB phase is the presence of A in (4.4a) or (4.13), *which are classical equations*. The discussions in Sect. 4.6 on the variation of Λ are as relevant to the argument as are the above considerations of the general structure of quantum theory.

5.7.1 An Alternative Form of Basic Equations

In 1962 DeWitt [5.18] showed how the basic wave equation for an electron in a given electromagnetic field could be written in terms of the fields $F_{\mu\nu}(x)$ instead of the potentials $A_\mu(x)$. This new formulation was obtained by using a gauge transformation.

In DeWitt's formulation the field strengths appear non-locally (in line integrals), whereas in the usual formulations the potentials $A_\mu(x)$ appear locally. Belinfante [5.19] and S. M. Roy [5.20] also developed this method. Roy showed that DeWitt's gauge transformation is only possible from a physical point of view provided further conditions (and restrictions) are applied that guarantee the single-valuedness of the wave function.

U. Klein [5.21] examined the gauge transformation for the case of a multiply connected space R, such as the region outside an infinite solenoid. He showed that it is not possible to have $F_{\mu\nu}$ vanishing over the region R and also have the (new) potential A'_μ going to zero over R, without violating the condition for single-valuedness.

Thus in particular the vector potential A'_μ of an infinitely long solenoid cannot be made to obey $A'_\mu \equiv 0$ outside the solenoid.

Hence DeWitt's device will not provide a way of denying the existence of the AB effect by asserting that only the field $F_{\mu\nu}$ matters in the case of the AB effect.

In Sects. 4.1.2 and 4.3.3 of Ref. [5.17] and in Sect. IV C of Ref. [5.9], there are surveys of these arguments.

5.7.2 An Argument of Bocchieri and Loinger

Another mathematical situation arises in the criticism of the AB effect by Bocchieri and Loinger [5.22] (see also Ref. [5.23]). Consider an infinite cylindrical solenoid as in Fig. 2.1; the vector potential in the region $\rho > \rho_C$ outside the solenoid is given by (2.3), so there is only one component A_θ and

$$A_\theta = \frac{\Phi}{2\pi\rho} \ , \qquad \text{for } \rho \geq \rho_C \ . \tag{5.9}$$

A gauge transformation is of the form

$$\boldsymbol{A}(\boldsymbol{x}) \rightarrow \boldsymbol{A}'(\boldsymbol{x}) = \boldsymbol{A}(\boldsymbol{x}) + \boldsymbol{\nabla}\chi(\boldsymbol{x}) \ . \tag{5.10}$$

On choosing

$$\chi = -\Phi\frac{\theta}{2\pi} \tag{5.11}$$

the vector potential in $\rho \geq \rho_C$ has only the component A'_θ and moreover

$$A'_\theta = 0 \qquad \text{for } \rho \geq \rho_C \ .$$

Hence one might claim that there is no AB effect [5.23].

In order to see what is wrong with this argument, it is useful to remember the basic idea of the gauge transformation in the case of wave mechanics. It is to preserve the matrix elements of the operator

$$\boldsymbol{p} + e\boldsymbol{A} \ .$$

Thus in addition to (5.10) we have to change the wave function according to

$$\psi \rightarrow \psi' = \exp\{-ie\chi(\boldsymbol{x})/\hbar\}\psi \ . \tag{5.10a}$$

Thus the electron probability destribution $|\psi(\boldsymbol{x})|^2$ is unaltered by the gauge transformation.

However it is clear that the change of gauge given by (5.11) has altered the eigenvalue of the operator

$$L_z = \frac{\hbar}{\mathrm{i}}\frac{\partial}{\partial\theta}$$

which was defined in (2.13a). It has added the angular momentum component

$$+ \hbar\Phi/\phi_e \tag{5.11a}$$

to the eigenvalue of L_z.

This gives the operator Λ_z, defined by (2.13), as the total component of the electron's angular momentum about the symmetry axis O_z.

5.7.3 Conclusion

The argument in Refs. [5.22], [5.23] uses a gauge transformation (5.11), that – on applying gauge invariance consistently – leads to the operator Λ_z which gives the electron's physical angular momentum about O_z, as is shown in Sect. 2.4. Also, as was seen in Chap. 2, the AB effect is still present in this approach. Nothing physical bas been eliminated by the gauge transformation. [Equations (5.10), (5.11) only give a flux transformation, as in Sect. 2.5, for the special case that $\Phi = (M' - M)\phi_e$ where M, M' are integers.]

5.8 The Inaccessible Field and Fringe Fields

In Sect. 5.7 we have just mentioned the mathematical difficulties that stand in the way of eliminating the effect of the inaccessible field in the original AB experiments. It should be emphasised that numerous authors have criticised the papers of Aharonov and Bohm [5.10], [5.15], [5.24] on the general ground that "the electron in Fig. 2.3 (Sect. 2.4) is not passing through an electromagnetic field, and therefore there cannot be an AB effect".

This is an important argument, and it would be true in classical physics that the inaccessible stationary magnetic field (or flux) cannot give rise to any force on the electron.

But even this classical argument is incorrect in the case that the inaccessible flux alters with time. Then electromagnetic induction causes a force on the electron, as is worked out explicitly in Sect. 2.7. Equation (2.20a) (in Sect. 2.7) shows that a small change $\delta\Phi$ in the inaccessible flux causes an alteration in the angular momentum Λ of the electron around the axis of cylindrical symmetry by the amount

$$\delta\Lambda = \frac{e}{2\pi}\delta\Phi \ . \tag{5.12}$$

Presumably it is only a matter of opinion, or of early education, whether the fact that Faraday's electromagnetic induction arises from the inaccessible flux is any less surprising, or more surprising, than the AB effect.

Equation (2.20a) is a classical result, but it also applies to a circulating wave packet in quantum theory. By the quantum theory adiabatic principle (cf. Sect. 8.8 and Refs. [8.10], [8.11]) the quantum number of the angular momentum about the axis of symmetry will be unaltered provided the change $\delta\Phi$ in the (inaccessible) flux is made sufficiently slowly.

The quantum theory formula (2.13) now gives the same result as (2.20a) on taking the expectation value with respect to the circulating wave packet.

The AB effect is an assertion about the phase of the (circulating) wave function in the case of an inaccessible flux that is unaltered in time, *relative to* the phase when there is zero flux. Thus the AB effect is not an absolute, but a relative effect. In order to detect it, a change in flux is needed, and at this point in the argument we have to remember Faraday's induction.

In Sects. 2.7, 2.8 it was explained how (2.20a) is quite consistent with the AB effect. Another way to state the conclusion consists in sharpening the precision of the language used. We can have inaccessible induction fields, but a change in the inaccessible flux will not itself be inaccessible. Critics often appear to have forgotten this basic property.

In Refs. [5.22], [5.23] authors have used "inaccessible field" or similar arguments in order to deny that the AB effect can occur.

Further discussion of this and closely related topics is given by Greenberger [5.25], and Greenberger and Overhauser [5.26].

5.8.1 Arguments Involving Fringe Fields

Various authors argue that the AB effect demonstrated in many experiments is, partly or wholly, the result of fringe induction fields acting on the electron by the Lorentz force; see for example [5.20], [5.22], [5.23], [5.27]. One significance of such arguments is that, if true, they contradict the assertion of Aharonov and Bohm [5.10], [5.15] that the AB phase is solely due to the potential $A_\mu(x)$.

Clearly these arguments are in principle distinct from the discussion about the inaccessible induction or flux that we have just examined.

The Chambers–Pryce effect that was considered in Sect. 5.6, and at the start of Sect. 5.7 above, is a simple case of such a "fringe-field" effect.

Bocchieri et al. [5.22], [5.28] also considered an experiment of Boersch et al. [5.29] that is of the biprism type in Fig. 5.1. The apparatus has a small cylinder to which a permalloy strip is attached for half of its length. The cylinder lies along the Oz direction (cf. Fig. 5.1) and the strip ends at $z = 0$. The strip is in the shadow of the cylinder (see also Sect. IIIC of Ref. [5.9] for further details).

The interference pattern in the region of $z = 0$ is shown in Fig. 5.6. On passing through $z = 0$ the *envelope* is shifted to the right, while the *fringes themselves* shift leftwards relative to the envelope. The shift is proportional to the flux in the strip.

Near the end of the strip there is a component of induction B along Oz. It is of the form (cf. Ref. [5.9])

$$B_z = (\text{const.})\frac{z}{(z^2 + \rho^2)^{3/2}} \tag{5.13}$$

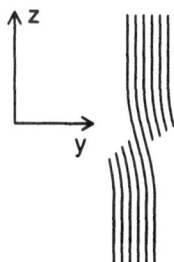

Fig. 5.6. Interference pattern near the end ($z = 0$) of the permalloy strip. The axes are as in Fig. 5.1

where ρ is the distance between the observation position and the axis of the (narrow) strip. The electrons move parallel to Ox, and the Lorentz force acting on them due to the induction B_z (cf. Fig. 5.1) lies along Oy.

The physics of the phenomena seen in Fig. 5.6 is therefore that near $z = 0$ the induction component B_z shifts the envelope of the interference pattern in the direction Oy, the sign of the shift being reversed at $z = 0$.

The abrupt change of flux at $z = 0$ causes the abrupt (backward) shift of the individual fringes across the envelope in the region of $z = 0$. This is an example, in strong from, of the Chambers tilted fringes (Sect. 5.6). This experiment was used successfully [5.29] to determine the basic constant ϕ_e.

Further details of the debates about the causes of various types of fringes and shifts in a number of experiments are found in Sect. 4.3.3 of [5.17].

5.9 Summary on "Potentials Versus Fields"

We can describe the various phenomena systematically as follows. In the ideal case of an infinite circular solenoid, the use of the potential A_θ of (2.3) gives the easiest description of the AB effect.

In the case of *changes* from this simple example, electromagnetic fields begin to play a role. There are several types:

(i) On altering the flux in the solenoid with time, the change in the AB effect is described by using electromagnetic induction, as discussed earlier in this chapter, and in Sects. 2.7, 2.8.

(ii) On altering the radius of the solenoid so that it becomes tapered, the Chambers–Pryce effect explains the changed fringe pattern on using the Lorentz force due to the B_ρ (radial) fringe field.

(iii) On extinguishing the flux in one half of the solenoid, the longitudinal induction B_z of (5.13) that exists near the "break" explains the new fringe pattern on using the Lorentz force. At the same time the discontinuous flux describes the strong tilting of the fringes at the "break" as a rather extreme example of (ii) above.

(iv) In the case of the generalized AB effect of Chap. 3, (namely motion in a uniform induction field, in the hollow axial symmetric field, or even in more general cases), there does not appear to be any useful purpose in arguing that potentials are more important than electromagnetic forces.

Conclusion

Thus for an infinite circular solenoid the vector potential A_θ of (2.3) has to be used in order to determine the AB effect.

From this result, on using electromagnetic fields, as in (i), (ii), (iii) above, the AB effect can be deduced for other infinite circular solenoids with different flux, for tapered solenoids, and for infinite solenoids in which the flux abruptly alters at a transverse plane.

For sources of flux with different topology, e. g. a toroidal solenoid, other methods would have to be used, and we do not discuss such cases.

This picture makes it difficult to assert that the potential $A_\mu(x)$ alone is the cause of the original AB effect. However, the potential A_θ (2.3) is apparently an essential starting point in calculations.

In the cases of the generalized AB effect in Chap. 3, mentioned in (iv) above, the potential $A_\mu(x)$ of (2.3) need not be used.

Return Flux etc. For the practical finite length circular solenoid the electromagnetic field is always present outside the solenoid. This case has been discussed by Lipkin [5.30] and by Roy [5.20] among others. Roy suggests that the observed effects, in the case of finite solenoids, can be due to the non-zero field outside the solenoid. Lipkin gives strong arguments against Roy's suggestion. Lipkin points out that the flux in the region actually traversed by the electrons does not determine the *flux* inside the (finite) solenoid [5.30].

A toroid solenoid is a good example of a finite system where the magnetic field in a given region cannot be determined by the field strength outside that region.

Early discussions and controversies about the AB effect, up to 1970, were surveyed by Erlichson [5.31] in detail.

5.10 Aharonov–Bohm Scattering

This two-dimensional scattering is a theoretical phenomenon that is related to the AB effect itself. Unfortunately in discussions about it, there has at times been confusion about mathematical features. Fuller details are given in Appendix G below, and here only a summary is provided.

An electron moves perpendicular to, and near, a magnetic fibre that lies along O_z and has flux Φ. Ideally the fibre has zero radius. The electron has a wave function $\psi(\rho, \theta)$ obeying

$$\frac{\partial^2 \psi}{\partial \rho^2} + \frac{1}{\rho}\frac{\partial \psi}{\partial \rho} + \left\{ k^2 + \frac{1}{\rho^2}\left(\frac{\partial}{\partial \theta} + i\alpha \right)^2 \right\} \psi = 0 \ . \tag{5.14}$$

ρ, θ are polar coordinates and motion along O_z is ignored. The vector potential is A_θ as given by (2.3) and

$$\alpha = \Phi/\phi_e \ . \tag{5.15}$$

In general α is not an integer.

It is obvious that an exact solution of (5.14) is

$$\psi_B(\rho, \theta) = \exp\{i(k\rho\cos\theta - \alpha\theta)\} \ . \tag{5.16}$$

As is seen in Appendix G, this wave function describes an electron moving parallel to the O_x axis with a constant velocity $v_x = \hbar k/m$.

For $\alpha = 0$ we have another wave function,

$$\psi_0(\rho, \theta) = \exp\{ik\rho \cos \theta\} \quad , \tag{5.17}$$

which gives plane motion parallel to O_x. The wave function $\psi_B(\rho, \theta)$ gives the same probability density $|\psi_B|^2$ as does $\psi_0(\rho, \theta)$. Therefore $\psi_B(\rho, \theta)$ gives no scattering. However $\psi_B(\rho, \theta)$ differs from $\psi_0(\rho, \theta)$ by a pure phase factor $\exp(-i\alpha\theta)$, and this is similar to the relation between $\psi_A(x)$ and $\psi_0(x)$ in Sect. 2.10 in the original AB effect. As was discussed in Sect. 2.10 there is no problem with single-valuedness of $\psi_B(\rho, \theta)$ as θ varies.

We can indeed cite $\psi_B(\rho, \theta)$ as a good example of the AB effect. A phase change relative to $\psi_0(\rho, \theta)$ is detectable.

Aharonov and Bohm [5.24] however imposed a further condition on the solution of (5.14), namely that

$$|\psi(\rho, \theta)| \to 0 \quad \text{as } \rho \to 0 \quad . \tag{5.18}$$

This additional condition provides a well-defined problem. As is shown in Appendix G, the general solution of (5.14) can now be written in the form

$$\psi_{AB}(\rho, \theta) = \sum_{m=-\infty}^{\infty} a_m J_{|m+\alpha|}(k\rho) \exp(im\theta) \tag{5.19}$$

where a_m are arbitrary constants. Each term on the right of (5.19) will vanish as $\rho \to 0$, provided α is not an integer (or zero). In this way (5.18) is obeyed.

On the other hand $\psi_B(\rho, \theta)$ [on using (G.8) in Appendix G] can be written as the well-behaved series

$$\psi_B(\rho, \theta) = \sum_{m=-\infty}^{\infty} i^m J_m(k\rho) \exp\{i(m - \alpha)\theta\} \quad . \tag{5.20}$$

This solution of (5.14) tends to the value $\exp(-i\alpha\theta)$ as $\rho \to 0$. It follows that unless α is an integer or zero, $\psi_B(\rho, \theta)$ cannot be written in the form (5.19).

Aharonov and Bohm [5.24] used the particular solution of (5.14)

$$\psi_{AB}^{(P)}(\rho, \theta) = \sum_{m=-\infty}^{\infty} i^{|m+\alpha|} J_{|m+\alpha|}(k\rho) \exp(im\theta) \quad . \tag{5.21}$$

If α is zero or an integer, $\psi_{AB}^{(P)}(\rho, \theta)$ is just $\psi_B(\rho, \theta)$ of (5.16). This can be seen by using (5.20) and (G.8a). For α not zero or not an integer, $\psi_{AB}^{(P)}(\rho\theta)$ obeys (5.18).

Aharonov and Bohm [5.24] show that on subtracting from (5.21) the function $\psi_B(\rho, \theta)$ of (5.16), the remainder behaves for large $(k\rho)$ as an outgoing wave; cf. (G.7a). This yields the scattering cross-section σ where

$$\sigma \, d\theta \simeq \frac{\lambda}{4\pi^2} \frac{\sin^2(\pi\alpha)}{\sin^2(\theta/2)} d\theta \quad . \tag{5.21a}$$

Again, $\sigma \, d\theta$ vanishes if α is zero or an integer.

The cause of this scattering is the imposition of condition (5.18) on $\psi_{AB}^{(P)}(\rho, \theta)$, plus the effect of the flux $\varPhi$ in the line O_z. This may seem strange since (in principle) there is no induction B outside the line O_z. So we are back to a familiar problem; namely what effect deflects the electron to give rise to the cross-section $\sigma\, d\theta$ of (5.21a)?

Notice that if we let $\hbar$ vanish, then λ will vanish (for the same momentum of the electron); hence $\sigma\, d\theta$ is a quantum effect that vanishes as $\hbar \to 0$. In Appendix G it is discussed how the boundary condition (as $\rho \to 0$) on the wave function could cause a surface effect that leads to $\sigma\, d\theta$ of (5.21a).

Details of the mathematical properties are in Appendix G, together with some comments on the nature of "two-dimensional scattering". See pp. 367–378 of Ref. [5.9] for a detailed discussion of the formalism.

A discussion of the question of the return flux in the AB scattering problem is to be found on p. 18 of Ref. [5.17]. The calculations in Refs. [5.32], [5.33] show that objections to the AB effect based on the return flux are irrelevant. Moreover the need for currents at infinite distance can be avoided by using instead a finite toroidal solenoid (cf. Appendix C).

5.11 Experiments with Superconductors

In recent years a number of elegant experiments have used electron holography in order to detect the wave pattern around a magnetised torus covered by a layer of superconducting material [5.17], [5.34]–[5.36]. A sketch of a typical torus is shown in Fig. 5.7. The cross-section of the magnet ring is of the order of 1–2 μm, and the thickness of the covering superconductor layer is of the order of 0.3 μm, or more. The strength of the induction in the magnet ring can be varied, and the shift in the interference fringes is clearly observed.

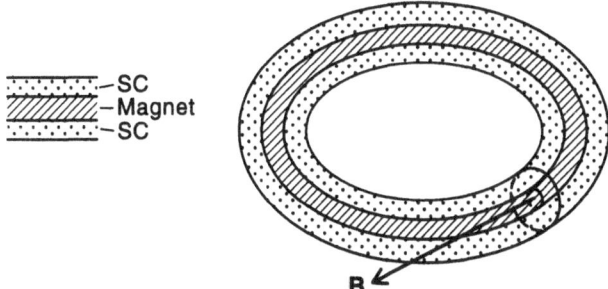

Fig. 5.7. Hatched region: magnetised torus. Dotted region: susperconductor (S.C.) covering. A typical cross section and its induction field B are shown

A practical use of electron holographic interference microscopy for investigating magnetic domain walls is given in [5.37]. This also makes use of the phase factor $U(2\pi)$ of (2.7a).

5.11.1 A Basic Equation

In order to understand the physical situation it is convenient to look at an infinite circular cylindrical magnet of radius R as in Fig. 5.8. Cylindrical coordinates (z, ρ, θ) are used, with O_z along the axis of the magnet. The induction B is along O_z and is uniform over the cylinder. The magnet is covered with a cylindrical sleeve of superconducting material which has inner and outer radii R and R', respectively.

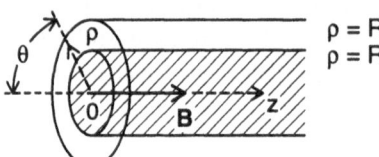

$\rho = R'$
$\rho = R$

Fig. 5.8. Infinite magnet $0 \le \rho \le R$ with uniform induction B covered with an infinite sleeve $R < \rho < R'$ of superconducting material

The induction in the magnet causes a current $j = (0, 0, j_\theta(\rho))$ in the superconductor and, as we shall see, there is also a field $B = (B_z(\rho), 0, 0)$ in the superconductor, especially near the inner surface $\rho = R$. London's penetration parameter is λ_L.

In the superconductor a macroscopic wave function $\psi(x)$ describes the coherent behaviour of the Cooper pairs (which have charge q and mass m). The long-range order is described by the formula

$$\psi(x) = (n_C)^{1/2} \exp(iS(x)/\hbar) \; , \tag{5.22}$$

where n_C is the number density of the Cooper pairs and $S(x)$ is a smooth function. (See for example Sect. 2.3 of Ref. [5.38] for a discussion.)

The current density, in notation consistant with (2.2) and Sect. 2.5 above, is

$$j = (iq\hbar/2m)\{\psi(\nabla\psi^*) - \psi^*(\nabla\psi)\} - (q^2/m)A\psi^*\psi \; . \tag{5.23}$$

Substituting (5.22) gives the basic equation

$$\nabla S = qA + \frac{m}{qn_C}j \; . \tag{5.23a}$$

It was pointed out years ago [5.39], [5.40] that (5.23a) is closely related to the AB effect. The last quantity on the right of (5.23a) is seen to be

$$mv$$

where v is an average velocity of a single Cooper pair. In the case of the superconducting sleeve in Fig. 5.8, we can integrate (5.23a) around a closed path which surrounds the magnet and remains in the superconductor. This gives

$$\phi S] = q \oint \boldsymbol{A} \cdot d\boldsymbol{s} + m \oint \boldsymbol{v} \cdot d\boldsymbol{s} \quad . \tag{5.23b}$$

In the last term in (5.23b) only the component v_θ contributes, so we have

$$\phi S] = q \oint \boldsymbol{A} \cdot d\boldsymbol{s} + \oint \Lambda \, d\theta \tag{5.24}$$

where Λ is the angular momentum about O_z. The term on the left is the change in S in the circuit.

Equations (5.23b) and (5.24) would follow directly from the classical equations (4.4), (4.4a) in Sect. 1, provided that

$$K = \frac{\phi S]}{2\pi} \tag{5.24a}$$

is taken for the (adiabatic) action variable. (Here for the Cooper pair we use $q = -2e$, $m = 2m_e$, whereas in (4.4) etc. $q = -e$ was used for the electron.)

This demonstrates the close relation between (5.23a) and the AB effect.

5.11.2 Flux Quantization

Consider how $|\boldsymbol{j}|$ behaves on moving from the inner surface $\rho = R$ through the superconductor. If $|\boldsymbol{j}|$ decreases faster than ρ^{-1}, then Λ becomes negligible relative to the flux term in (5.24). So if the superconducting sleeve is sufficiently thick, there is a circle $\rho = \rho'$ in the superconductor, such that for $\rho \geq \rho'$, Eqs. (5.22), (5.24) require

$$\Phi(\rho) = \frac{nh}{q} \quad , \qquad (n = 0, \pm1, \pm2, \ldots) \quad . \tag{5.25}$$

Here $\Phi(\rho)$ is the flux through the circle of radius ρ. The integer n arises from requiring that $\psi(\boldsymbol{x})$ of (5.22) be single valued. It follows that

$$K = n\hbar \quad ;$$

this is similar to what we had in (4.12).

Equation (5.25) is the flux quantization relation of Byers and Yang [5.39], [5.41]. The derivation above is only a sketch of the arguments that are relevant, and we have concentrated on the basic mathematical structure. For details of the physical ideas, see for example Sect. 11.9 of [5.42]. For the case that the inner radius R, or the thickness of the sleeve $R' - R$, are not appreciably larger than London's penetration depth λ_L, (5.25) is modified; see [5.43] for details. For verification of flux quantization for BCS superconductors see [5.44], [5.45].

5.12 The Behaviour of $B_z(\rho)$ and $j_\theta(\rho)$

Details of the behaviour of B_z and j_θ in the sleeve $R \leq \rho \leq R'$ in Fig. 5.8, are important in understanding the derivation of the flux relation (5.25) and in appreciating the implications of the electron holography experiments.

Equation (5.23a) yields

$$B + \frac{m}{q^2 n_C} \operatorname{curl} j = 0 \quad , \tag{5.26a}$$

and Ampéres law can be written

$$\operatorname{curl} B = \mu \mu_0 j \quad . \tag{5.26b}$$

In the superconductor we put $\mu = 1$. Using the forms of B and j that were use in Sects. 5.11, it is easy to see that in $R \leq \rho < R'$, Eqs. (5.26a) and (5.26b) give

$$\frac{d^2 B_z}{d\rho^2} + \frac{1}{\rho} \frac{dB_z}{d\rho} - \frac{1}{\lambda_L^2} B_z = 0 \quad , \tag{5.27a}$$

$$\frac{d^2 j_\theta}{d\rho^2} + \frac{1}{\rho} \frac{dj_\theta}{d\rho} - \left(\frac{1}{\lambda_L^2} + \frac{1}{\rho^2} \right) j_\theta = 0 \quad . \tag{5.27b}$$

The relation (5.26b) is simply

$$j_\theta(\rho) = -\frac{1}{\mu_0} \frac{dB_z(\rho)}{d\rho} \quad , \tag{5.27c}$$

and

$$\lambda_L^{-2} = \frac{q^2 n_C \mu_0}{m} \quad .$$

Equations (5.27a), (5.27b) are satisfied by hyperbolic Bessel functions of order 0 and 1 respectively [5.46]. The solutions $I_0(\rho/\lambda_L)$, $I_1(\rho/\lambda_L)$ will increase rapidly with ρ, when ρ/λ_L exceeds about 2. For a sleeve whose thickness exceeds $2\lambda_L$, the solutions $K_0(\rho/\lambda_L)$ and $K_1(\rho/\lambda_L)$ must be used. For large $|z|$ the asymptotic forms are

$$K_0(z) \simeq \left(\frac{\pi}{2z} \right)^{1/2} e^{-z} \left\{ 1 - \frac{1}{8z} + O\left(|z|^{-2}\right) \right\}$$

$$K_1(z) \simeq \left(\frac{\pi}{2z} \right)^{1/2} e^{-z} \left\{ 1 + \frac{3}{8z} + O\left(|z|^{-2}\right) \right\} \quad .$$

For positive real z, with these sign conventions,

$$K_1(z) > K_0(z) > 0 \quad .$$

We shall use the relations

$$K_0'(z) = -K_1(z) \quad , \qquad \frac{d}{dz}\{zK_1(z)\} = -zK_0(z) \quad .$$

If the uniform field in the magnet ($0 \leq \rho \leq R$) is $B_z = B$, the solution of (5.27a) in the sleeve is

$$B_z(\rho) = BK_0\left(\frac{\rho}{\lambda_L}\right)\Big/K_0\left(\frac{R}{\lambda_L}\right) \tag{5.28}$$

and by (5.27c)

$$j_\theta(\rho) = (\lambda_L\mu_0)^{-1}BK_1\left(\frac{\rho}{\lambda_L}\right)\Big/K_0\left(\frac{R}{\lambda_L}\right) \quad . \tag{5.28a}$$

For $\rho \gg \lambda_L$,

$$j_\theta(\rho) \simeq (\lambda_L\mu_0)^{-1}B_z(\rho) \quad . \tag{5.28b}$$

Thus the system in Fig. 5.8 behaves like a solenoid, with the current j_θ confined to a layer of thickness λ_L or so, just outside $\rho = R$. In practice λ_L is small, e.g. for Nb, $\lambda_L = 0.1\,\mu$m.

The flux in the sleeve is

$$2\pi \int_R^{R'} d\rho\,\rho B_z(\rho) = 2\pi\lambda_L B\left\{RK_1\left(\frac{R}{\lambda_L}\right) - R'K_1\left(\frac{R'}{\lambda_L}\right)\right\}\Big/K_0\left(\frac{R}{\lambda_L}\right)$$

$$\simeq 2\pi\lambda_L RBK_1\left(\frac{R}{\lambda_L}\right)\Big/K_0\left(\frac{R}{\lambda_L}\right)$$

$$\simeq 2\pi\lambda_L RB \quad . \tag{5.29}$$

In this evaluation it is assumed that $R/\lambda_L \gtrsim 2$, and $R' - R \gtrsim 3\lambda_L$. The total flux is therefore

$$\Phi \simeq \pi R^2 B + 2\pi\lambda_L RB \quad . \tag{5.29a}$$

5.12.1 Result of the Experiments

The results of the electron holography experiments in which the part of an electron beam passing the torus in Fig. 5.7 interferes with the part that goes through the torus are given in [5.34]–[5.36]. When the superconducting sleeve is kept below the critical temperature T_C the interference pattern shows either no shift of the fringes, or else a shift of half the fringe separation Δ_f.

These result show two things: (a) the AB shift exists, and is clearly demonstrated; (b) the flux is quantized according to (5.25) with $|q| = 2e$.

If the flux in the toroidal magnet is of the order of $5\phi_e$, [5.35], then using the calculation for the covered cylinder of Fig. 5.8, it follows from (5.29a) that the flux Φ' in the superconductor itself is of the order of

$$\Phi' \simeq 10\frac{\lambda_L}{R}\phi_e \quad .$$

If R is of the order of $1\,\mu$m and λ_L is $0.1\,\mu$m (and such values have indeed been used), then $\Phi' \simeq \phi_e$.

In such a case the electron holography experiment may not be able to give much information about the flux in the magnet itself. Nevertheless, the experiment does demonstrate the AB effect, and the total flux is quantized as a consequence of (5.23a) and (5.25).

6. Accuracy of Flux Measurements

This chapter begins with a discussion of the significance of the various meanings of the uncertainty, or error, symbols Δ that are in use. The increased accuracy that can be achieved by repeating a measurement many times is examined, together with an example from a beam guidance system.

Heisenberg's result on the quantum theory limit on the accuracy of measuring magnetic flux, by using the Lorentz force, is examined. If is shown that the accuracy that can be achieved using the AB effect is much the same. Comparison is made with the (theoretical) limit on the flux accuracy measurement using SQUID and similar devices.

6.1 Uncertainty Notation

The quantum error symbol Δ has been, and is, used to indicate different quantities, so it is necessary to state briefly the notation to be used in this chapter. $\langle x \rangle$ is the expectation of a Hermitian operator x in a specified state. The variance $\Delta_v x$ is defined by

$$(\Delta_v x)^2 = \langle (x - \langle x \rangle)^2 \rangle \quad .$$

For Hermitian operators A, B, C obeying

$$[A, B] = iC$$

we have

$$\Delta_v A \Delta_v B \geq \frac{|\langle C \rangle|}{2} \quad . \tag{6.1}$$

In his book [6.1] Werner Heisenberg uses

$$\Delta x = 2^{1/2} \Delta_v x \tag{6.2}$$

for "the uncertainty in the knowledge of x" in the given state (cf. Eqs. (5), (7) in [6.1]). This yields the well-known relation

$$\Delta A \Delta B \geq |\langle C \rangle| \tag{6.2a}$$

in place of (6.1).

We could introduce a third estimate of uncertainty:

$$\Delta_{70} x = 2\Delta_v x \quad . \tag{6.3}$$

For Gaussian or similar distributions, there is about 70 % certainty that the measurement of x will lie in an interval $\Delta_{70}x$ centred on $\langle x \rangle$. This is a realistic way of measuring accuracy. It yields

$$\Delta_{70}A\Delta_{70}B \geq 2|\langle C \rangle| \ . \tag{6.3a}$$

Thus for a canonical conjugate pair p, q

$$\Delta_{70}p\Delta_{70}q \geq \frac{h}{\pi} \ . \tag{6.4}$$

The equality in (6.1), (6.2a), (6.3a), (6.4) holds only for the minimum uncertainty states. For a canonical conjugate pair these states give a Gaussian probability distribution in the wave packet. It is important in practice to realise that otherwise the error products can be appreciably larger than the minimum value. Numerous physicists have used the practical rule:

$$\Delta p\Delta q \geq h \ , \tag{6.5}$$

when not specifically dealing with the minimum uncertainty states.

6.2 Frauenhofer Diffraction Example

Consider the diffraction of a plane uniform monochromatic electromagnetic beam of wavelength λ, falling on a slit of width b (Fig. 6.1).

In the Frauenhofer region at angle θ, the intensity is

$$I(\theta) = I_0 \left(\frac{\sin \alpha}{\alpha} \right)^2 \tag{6.6}$$

with

$$\alpha = \frac{\pi b}{\lambda} \sin \theta \ . \tag{6.6a}$$

I_0 is the intensity at the point N (i.e. in the direction $\theta = 0$).

The intensity falls to half its maximum value I_0 at an angle

$$\theta_{1/2} = \pm 0.44\frac{\lambda}{b} \tag{6.6b}$$

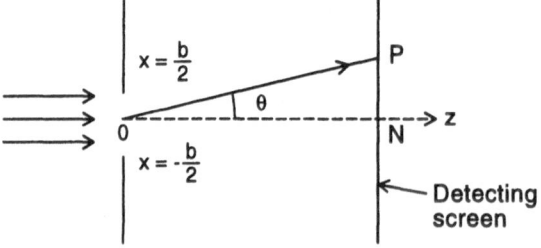

Fig. 6.1. Frauenhofer diffraction from a infinite slit $(-b/2 \leq x \leq b/2)$, $(-\infty < y < \infty)$ on a plane screen

on either side of the forward direction (in the case where λ/b is small). If the incident momentum of a photon is p, the spread in transverse momentum is taken to be

$$\Delta p_x \simeq 2p\theta_{1/2} \simeq 0.9\frac{h}{b} \quad . \tag{6.7}$$

It is reasonable to take the uncertainty in the position where the photon passes the plane $z = 0$ to be $\Delta x = b$. So (6.7) yields

$$\Delta p_x \Delta x \simeq 0.9h \quad . \tag{6.7a}$$

This argument supports the use of the practical rule of (6.5).

When the amplitude and phase of the wave falling on $z = 0$ are not uniform, the situation can be quite different from that in (6.6), (6.7). Such is the case with the coherent beam emerging from the resonant cavity (of length L) in a laser. In the lowest mode of the cavity, the radius of the radiation waist w_0 (inside the cavity), the radius of the exit spot w_s and the semi-angle of the beam divergence θ are, respectively

$$w_0 = \{L\lambda/2\pi\}^{1/2} \quad , \qquad w_s = 2^{1/2}w_0 \quad , \qquad \theta \simeq 1.6w_0/L \quad .$$

For $L = 0.3\,\mathrm{m}$, $\lambda = 0.1\,\mu\mathrm{m}$ the values are $w_s = 10^{-4}\,\mathrm{m}$, $\theta = 0.3 \times 10^{-3}\,\mathrm{rad}$. On using $\Delta p = 2\theta p$, $\Delta x = 2w_0$, $p = h/\lambda$, it is seen that (6.5) is again valid.

6.3 Increased Accuracy Using N Photons

Equation (6.6) gives the probability distribution for a single photon, in the case of a uniform incident field, and (6.7a) gives the error product for a single photon. The direction of the incident beam (i. e. $\theta = 0$ in Fig. 6.1) can be determined with much greater accuracy on using many photons.

This can be seen in the following crude way. The plot of $I(\theta)$ versus θ is almost flat near $\theta = 0$. Make the rough assumption that all the N photons that are used fall within the angle $2\theta_{1/2}$ about $\theta = 0$. The average angular separation between adjacent photons is

$$\bar{\delta} = 2\frac{\theta_{1/2}}{N} \quad .$$

The central direction can be determined from the distribution of photons by picking out the *median* photon (or the point mid-way between the two central photons). There are $N/2$ [or $(N-1)/2$] photons to the left of the median, but there is the Poisson fluctuation $(N/2)^{1/2}$ in their number. Thus the error in determining the central direction of the beam is

$$\left(\frac{N}{2}\right)^{1/2}\bar{\delta} = \left(\frac{N}{2}\right)^{-1/2}\theta_{1/2} \quad . \tag{6.8}$$

This situation can be examined in another fashion. If the uncertainty Δp_x of (6.7) in the transverse momentum of a photon were to represent a

Gaussian distribution with variance σ, then the mean transverse momentum of N photons:

$$\bar{p}_x = \frac{1}{N} \sum_{j=1}^{N} (p_x)_j \qquad\qquad (6.8a)$$

would have variance $N^{-1/2}\sigma$. [Here $(p_x)_j$ are the individual measurements.] The same result will hold for any symmetrical random distribution.

Hence the central direction of a diffraction pattern can be determined with high accuracy by using a large number N of photons. This situation is not to be confused with Rayleigh's criterion for the resolution of two objects on using a circular exit aperture of diameter b; that is the angle

$$\alpha = 1.22\frac{\lambda}{b} \quad ,$$

[cf. $2\theta_{1/2}$ from (6.6b) for the case of the slit].

It is necessary that the beam and the slit remain absolutely steady during the time required to collect N photons. This requirement is quite different from that in most gedanken experiments in which the product $\Delta A \Delta B$ is investigated. It should be noted that the method above does not use a classical limit in the strict sense. It merely uses the fact that $I(\theta)$ of (6.6) gives the probability distribution of a single photon.

An interesting historical example of the principle involved is the Lorenz system for aircraft guidance (Fig. 6.2). Two similar antennae have axes along slightly different horizontal directions and they transmit radio beams that have an appreciable overlap. The carrier wave of one beam is modulated to transmit dots, while the carrier of the other beam is modulated to transmit the complimentary dashes. An aircraft flying in either beam can use the dots or the dashes to move towards the central line OP. When it gets into the overlap region it moves towards the vertical plane in which the dots and dashes merge to give a continuous audio signal. The zone in which the two signals are thus detected to be of equal strength can easily be 10^{-2} to 10^{-3} times the width of either individual beam. The two beams will have different carrier frequencies, and it is the counting of dots and dashes that is important. For details see pp. 98–99 and pp. 123–4 of [6.2].

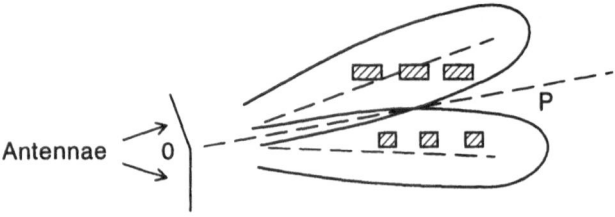

Antennae

Fig. 6.2. Lorenz system for aircraft guidance

We could use (6.6b) and (6.8) as a rough model. With $\lambda = 10\,\text{m}$, $b = 40\,\text{m}$, we have $\theta_{1/2} = \pm 5°$. Taking $N \simeq 2 \times 10^6$ in (6.8) yields a width of 30 m at 300 km. The Lorenz system is not a photon counting device, but it is a pulse strength measurer, and the basic principle is much the same as was used to get (6.8).

6.4 Heisenberg's Flux Measurement

Heisenberg estimated the limit on accuracy ΔB in measuring the induction B by using the force on a moving charge q:

$$\boldsymbol{F} = q(\boldsymbol{E} + \boldsymbol{v} \times \boldsymbol{B}) \ . \tag{6.9}$$

He wished to examine the accuracy in simultaneous measurements of B_z and E_x, in connection with the commutation relations of quantum electrodynamics. Our purpose is simpler and his scheme can be modified.

Let B be along O_z, and choose the test particle's velocity to the v along O_y. We do not need to assume any electric intensity E to be present, but for completeness we assume there is a component E_x, and that adds to the Lorentz force in (6.9). This effect is eliminated by sending through another identical particle of charge q with velocity $(-v)$. The difference in the behaviour of the two test particles will yield B_z [6.1]. The volume of space surveyed (which we call the designated space) is a rectangular box with edges d, L along O_x, O_y and a large depth along O_z. It is assumed that B and E vary very little over this box of space, or over the duration of the measurement. The length L is adjusted so that the particles do not bend far away from the direction O_y. The situation is sketched in Fig. 6.3.

A simple way to represent the motion is by taking the image in the plane Ozx of the path of particle 2 and plotting it with the actual path of particle 1, as in Fig. 6.4. The curve E is midway between paths 1 and 2, and it gives the orbit expected for $B = 0$.

Let us slopes of 1 and E differ by γ at $y = L$. From (6.9)

$$\gamma \simeq \frac{qLB}{p} \tag{6.9a}$$

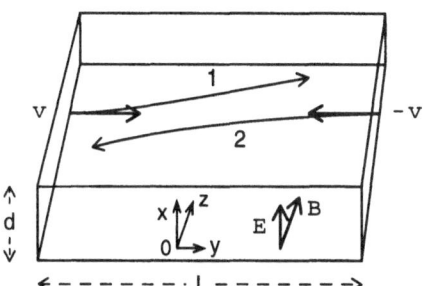

Fig. 6.3. Surveyed volume and fields E, B

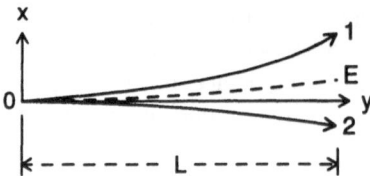

Fig. 6.4. Representation of the orbits in the Oxy plane

where p is the initial momentum along Oy. Both particles pass through a slit of width d to enter the designated space. As in Sect. 6.2, diffraction spreads each beam, and we can use (6.5) or (6.7a). The uncertainty of the momenta in the Ox direction is

$$\Delta p_x \gtrsim \frac{h}{d} \; , \tag{6.10}$$

and, for each beam, that gives an uncertainty in the slope at $y = L$ of

$$\Delta\gamma \gtrsim \frac{h}{pd} \; . \tag{6.10a}$$

This puts a limit ΔB on the accuracy of B when it is determined by (6.9a) which is given by

$$dL\,\Delta B \gtrsim \frac{h}{q} \; . \tag{6.10b}$$

In the notation of Chap. 2,

$$\Delta\Phi \gtrsim \phi_q \; , \qquad \left(\phi_q = \frac{h}{|q|} \right) \tag{6.11}$$

where $\Phi = dL\,B$ is the flux across the face Oxy of the designated space. (Heisenberg only needed $\Delta B_z \Delta E_x$, and he did not derive the form in (6.11) [6.1].) It is interesting to speculate whether the simple form in (6.11) might have some generality. We can specify it as the accuracy attainable when measuring the flux by using two identical objects of charges q. (As we are in effect using half of the difference in the slopes of curves O1 and O2 in Fig. 6.4 and the errors are independent, the results in (6.10a) and (6.11) overvalue the uncertainties ΔB or $\Delta\Phi$ by a factor $2^{1/2}$. For simplicity we shall ignore this point.)

This speculation is useful in considering the nature of the AB effect, and certain related properties of superconductors are relevant also.

6.4.1 Accuracy when Using 2N Test Particles

In Sect. 6.3 N photons were used to give a highly accurate determination of the axis of an electromagnetic beam (in the case of large N). In the same way in the gedanken experiment in Fig. 6.3 the uncertainty in each beam can be very much reduced by using the average in (6.8a), giving

$$\Delta \bar{p} \gtrsim \frac{h}{\mathrm{d}N^{1/2}}$$

in place of (6.10). Thus using N particles in each of beams 1 and 2 (and keeping the apparatus sufficiently still) we get

$$\Delta \Phi \gtrsim \phi_q N^{-1/2} \tag{6.11a}$$

for the quantum limitation on the accuracy of determining the flux Φ.

It follows from (6.10) that if the test charge q is replaced by a charge Nq, the uncertainty $\Delta \Phi$ is reduced by a factor N. This can be looked on as a coherent effect. Equation (6.11a) shows that on going from two particles to $2N$ particles, the reduction factor is $N^{-1/2}$. This is a form of incoherent effect in measurement.

6.5 Interference Measurements

In many cases interference will greatly improve the accuracy of a measurement. However, an interference pattern requires the passage of numerous particles, or photons, so there may be some relation between the accuracy in interference experiments and the devices in Sect. 6.3 and Sect. 6.4 for improving accuracy.

The probability, or density distribution $J(y)\,dy$ for a single fringe is shown in Fig. 6.5. The fringe spacing is Δ_{f} and

$$J(y) = \left(\Delta_{\mathrm{f}}^{-1}\right)\left\{1 + \cos(2\pi y/\Delta_{\mathrm{f}})\right\} \ . \tag{6.12}$$

The normalization is

$$\int_{-\Delta_f g/2}^{\Delta_f/2} J(y)\,dy = 1 \ ;$$

and 82 % of the events lie in a range $\Delta_{\mathrm{f}}/2$ centred on $y = 0$. For simplicity of argument we assume that the fringe is one-dimensional only.

Let N particles lie in this fringe. Their average separation near $y = 0$ is approximately

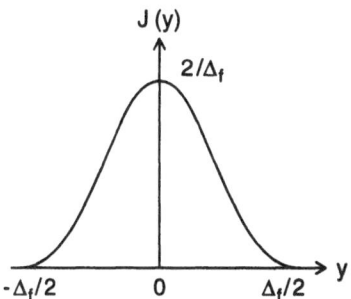

Fig. 6.5. Distribution $J(y)$ for a single fringe

$$\frac{\Delta_f}{1.6N} \quad.$$

The uncertainty as to which particle is central is $(N/2)^{1/2}$, so the position of the centre of the fringe is uncertain by

$$\frac{\Delta_f}{2.3N^{1/2}} \quad. \tag{6.12a}$$

An interference pattern will consist of a number of fringes. The number of fringes in a practical pattern is determined by the coherence length and electro-optical factors, as mentioned in Sects. 5.4, 5.5 above.

It appears that the fringes in Fig. 6.6 are equidistant, in which case the second and third fringes should be moved to the left by Δ_f and $2\Delta_f$ respectively. The correct value of Δ_f produces the greatest concentration of points in Fig. 6.7. This could be judged by making a hologram of the number of points in a standard small interval. If the separation of fringes is not uniform (but not random), a similar procedure could be devised, provided there are sufficient fringes.

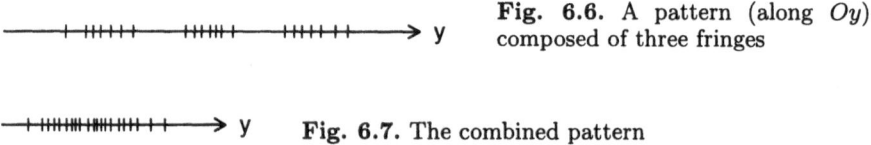

Fig. 6.6. A pattern (along Oy) composed of three fringes

Fig. 6.7. The combined pattern

If the total number of points in Fig. 6.7 is N, then the accuracy of determining the position of the pattern is given by (6.12a).

If there are F fringes, then the quantity

$$\frac{\Delta_f}{2.3}\left(\frac{F}{N}\right)^{1/2}$$

is the accuracy of determining the position of any one fringe. The mean of these F positions has the accuracy in (6.12a).

6.5.1 Application to the Static AB Effect

We refer to an experiment of the type in Sect. 5.1 in which an interference pattern with N particles is obtained while the flux in the solenoid is Φ_1. Then the beam of particles is switched off while the flux is altered to Φ_2; after this another interference pattern with N particles is obtained. By (5.2), the positions of the two interference patterns differ by

$$\Delta_f \frac{\Phi_2 - \Phi_1}{\phi_e}(\mathrm{mod}\,\Delta_f) \quad. \tag{6.13}$$

By (6.12a) we estimate that the smallest shift Δy between the two patterns that can be detected is of the order of

$$\Delta y \gtrsim 1.5 \Delta_f N^{-1/2} \quad .$$

Therefore

$$\Delta \Phi \gtrsim 1.5 \phi_e N^{-1/2} \tag{6.13a}$$

is an estimate of the accuracy of determining a change of flux on using N electrons in each (linear) interference pattern. The one-dimensional pattern has been used here because in principle it is possible to ignore the spread of the points in the directions Oz (cf. Fig. 5.1).

In practice the interference patterns are two-dimensional in the plane Oyz of Fig. 5.1. This requires many more particles to show fringes than in the one-dimensional model above. For example Ref. [6.3] shows a case in which five fringes are not clear to the eye until the order of 3 000 electrons have been recorded.

6.5.2 The AB Effect Versus Heisenberg's Experiment

Equation (6.13a) shows that, in ideal experiments, the accuracy of detecting the flux Φ by the AB effect is about the same as the accuracy in Heisenberg's gedanken experiment (6.11a), provided the number of electrons used in the two cases is respectively $2N$ and N. This is a further useful indication that the AB effect is consistent with related physical phenomena.

There is a difference when only one, or a few, electrons are used. In that situation no interference pattern can be detected. If someone wishes to assert that the AB effect does not exist for a single electron (or even for a few electrons) he or she cannot be proved wrong. On the other hand, the statement would appear to be at the most purely of philosophical interest. Notice that in Heisenberg's experiment, even the passage of two particles gives useful information about the flux Φ.

6.6 Other Considerations on Flux Measurement

In Chap. 3 the circular motion of an electron, or a particle of charge q, in a uniform induction B was discussed. It may appear that precise measurement of the cyclotron frequency ω_B of (3.5a) would give an accurate measurement of B, and thereby defeat the conjectured limitation in (6.11).

The wave function ψ is given in (3.3). [The conditions under which the dependence of ψ on the azimuthal angle θ can be of the simple nature in (3.3) are discussed in Sect. 3.3 (iii).] In this form a single eigenstate gives no azimuthal structure, since $|\psi|^2$ is independent of θ. In order to get a classical picture of a particle circling around, a set of (adjacent) values of the quantum number M is required. By (3.4a) this corresponds to a spread $\Delta \Phi$ in flux, where

$$\Delta \Phi = \phi_q \Delta M \quad ,$$

ΔM being some measure of the spread in M. The smallest practical value to give any picture of a particle would have $\Delta M \simeq 3$. The method does not improve on (6.11).

6.6.1 Superconductor Physics

The devices that use the Josephson effect [6.4] can provide very precise measurements of flux. A thorough discussion on SQUIDS (superconducting quantum interference devices) is given in [6.5].

A dc SQUID contains two Josephson junctions connected in parallel on a superconducting loop, where there is a static bias current. On changing the magnetic flux through the loop, the voltage across the loop oscillates with the period ϕ_q (where $q = 2e$). An electronic circuit converts the small voltage alteration into a current that passes through a coil, which, in turn, is coupled to the SQUID and cancels out the original change in flux. The small current can be measured [6.5]–[6.7].

A typical low T_C device of this type is very sensitive (to flux) and has a noise level of $10^{-6}\phi_q\,\mathrm{Hz}^{-1/2}$ (i. e. $10^{-6}\phi_q$ in a 1 Hz band width). Notice that $\phi_q(q = 2e) = 2.07 \times 10^{-7}\,\mathrm{Gauss\,cm}^2$.

The superconductor wave function (5.22) is a macroscopic quantum variable and the average density of Cooper pairs n_C is large (say $10^{24}\,\mathrm{m}^{-3}$) so considerable amounts of charge will act coherently in superconducting devices.

We wish to have an estimate of any essential quantum limit on the accuracy of flux measurements by superconducting devices. In practice the limit may well be determined by thermal noise, but that does not concern us here (see however [6.8], [6.9]).

In a SQUID or similar device [6.10] two currents that pass through Josephson junctions cause interference so that a phase difference

$$\beta = \beta_1 - \beta_2 \tag{6.14}$$

is produced and can be measured

The important question can now be put as follows: "Is there a quantum theory limit on the accuracy of measuring β, and if so, what is its form?" [The relation of the β_i, $(i = 1, 2)$ to the wave functions can be seen from (5.22) above.]

6.7 The Number–Phase Relation

It appears that the relation

$$\Delta n \Delta \phi \gtrsim 2\pi \tag{6.15}$$

[where the symbol Δ is used in the practical sense of (6.5)] may give the limitation we want. Here we are considering, for example, a quasi-monochromatic electromagnetic wave packet which has mean frequency ω; n is the number of photons present and ϕ is the phase at the centre of the wave packet.

There are difficulties about the status of (6.15), since n and ϕ are not, in general, a canonical conjugate pair of Hermitian operators. A survey of the awkward features is given in [6.11]. The relation

$$\Delta E \Delta t \gtrsim h$$

(which is valid for a wave packet) has been used to support the validity of (6.15) [6.12]. Dividing ΔE by ω and mutliplying Δt by ω would appear to give what we want, but one may prefer a firmer argument.

6.7.1 Coherent Radiation

A particularly good case in which to establish (6.15) is in the expectation values between optical coherent states. This is a basic part of physics and the underlying theory is precise and simple. We shall return at the start of Sect. 6.8 to comment on the application to the BCS theory of superconductivity.

A coherent state vector for a particular electromagnetic mode can be written as

$$|v\rangle = \exp(-|v|^2/2) \sum_{n=0}^{\infty} \frac{v^n}{(n!)^{1/2}} |n\rangle \quad , \tag{6.16}$$

where

$$v = |v| \exp(\mathrm{i}f)$$

is any complex number and $|n\rangle$ is the (Fock) state with photon number n for this mode. $|v\rangle$ obeys

$$a|v\rangle = v|v\rangle \quad ,$$

a being the operator for destroying a photon of this type having frequency ω. Also

$$\tilde{a}a|n\rangle = n|n\rangle \quad , \qquad (n = 0, 1, 2, \ldots)$$

is the number operator relation.

An important canonical conjugate pair of operators is

$$Q(t) = (2\omega)^{-1/2}\{a\exp(-\mathrm{i}\omega t) + \tilde{a}\exp(\mathrm{i}\omega t)\}$$
$$P(t) = (\omega/2)^{1/2}\{a\exp(-\mathrm{i}\omega t) - \tilde{a}\exp(\mathrm{i}\omega t)\}/\mathrm{i} \quad .$$

They obey

$$[P(t), Q(t)] = \mathrm{i}^{-1} \quad . \tag{6.17}$$

These operators describe the harmonic motion of the radiation oscillator for the mode concerned. Their expectation values in state $|v\rangle$ are simply

$$\langle v|Q(t)|v \rangle = (2/\omega)^{1/2}|v| \cos \phi \ , \tag{6.17a}$$

$$\langle v|P(t)|v \rangle = -(2\omega)^{1/2}|v| \sin \phi \ , \tag{6.17b}$$

where

$$\phi = \omega t - f$$

is the oscillator phase. In state $|v\rangle$ the expectation value of the photon number is

$$N = \langle v|\tilde{a}a|v \rangle = |v|^2 \ . \tag{6.17c}$$

We now use Q and P to denote the expectation values of $Q(t)$ and $P(t)$ in state $|v\rangle$. From (6.1), (6.17) we have

$$\Delta_v Q \Delta_v P \geq \tfrac{1}{2} \ . \tag{6.18}$$

Equations (6.17a, b) relate Q, P to $|v|$ and ϕ. It is easy to see, using the Jacobian, that any small changes δQ, δP will obey

$$\delta Q \delta P = 2|v| \, \delta |v| \, \delta \phi \ .$$

Therefore the uncertainties in (6.18) give rise to the relation

$$\Delta_v N \Delta_v \phi \geq \tfrac{1}{2} \ . \tag{6.19}$$

6.8 Application to Superconductors

The importance of the uncertainty relation (6.15) between the number of Cooper pairs in a BCS superconductor and the phase $S(x)/\hbar$ of the super-conducting wave function $\psi(x)$ in (5.22) has been emphasised by Tinkham [6.13], while Anderson [6.14], [6.15] has discussed the application of this un-certainty relation (6.15) to the Josephson effect and related topics.

In a dc SQUID (cf. the brief description in Sect. 6.6 above) employing two Josephson junctions in an ingenious fashion, the phase β in (6.14) can be related to the flux Φ through a given area, and this is arranged so that the flux quantization rule (5.25) does not apply. If the device is operated so that the current density term j in (5.23a) is *negligible* [6.16], then the change in the phase β on passing around a circuit is proportional to the integral of A around the circuit. Thus

$$\beta = \frac{q}{\hbar}\Phi \quad (q = 2e) \ , \tag{6.20}$$

and Φ can be measured.

In the case that the term in j in (5.23a) is not negligible, the simple relation (6.20) does not hold; the general result is closely related to the AB relation, as is discussed after (5.23b), (5.24) in Sect. 5.11 above. The classical equivalent is the action equation (4.4a).

The superconducting material consists of a very large number of Cooper pairs in a state that is similar to the optical coherent state $|v\rangle$ (for large $|v|$)

that we have just discussed [6.17]. In particular (6.19) should apply where N is the number of Cooper pairs and β replaces ϕ (see e. g. Sects. 7.2, 7.3 in [6.18]). Thus on going over to notation analogous to (6.5), we get

$$\Delta N \Delta \Phi \gtrsim \phi_{2e} \ . \tag{6.20a}$$

On writing $\Delta Q = 2e \Delta N$, we have

$$\Delta \Phi \gtrsim h/\Delta Q \ . \tag{6.21}$$

Since N is very large, ΔQ will be a fairly large charge, and high accuracy can be achieved. Equation (6.21) replaces Heisenberg's result (6.11) in the case of superconducting devices. The charge $|q|$ in (6.11) is now replaced by the uncertainty in charge ΔQ.

6.9 Experimental Verification of the Phase Uncertainty in a Superconductor

An experiment by Ellon et al. [6.19] gives evidence in support of (6.19) for the relation between the phase of the macroscopic wave function and the number of Cooper pairs.

Anderson [6.14] pointed out that for an isolated grain of superconductor, Eq. (6.19) implies that the phase of the wave function is quite uncertain. On the other hand, for two macroscopic superconductors that are connected by a thin insulating barrier, the phase difference between the two is well determined. Hence the number of Cooper pairs in each superconductor fluctuates considerably.

The experiment studies intermediate situations in which the uncertainties in n_C and in the phase are estimated to be comparable. Fabrication on a very small scale is required, and the resulting equipment is not simple. However, it is clear [6.19] that a non-trivial consequence of (6.19) is obeyed.

7. Phenomena Similar
to the Magnetic Aharonov–Bohm Effect

Various phenomena such as the electrical Aharonov–Bohm (EAB) effect, the Aharonov–Casher (AC) effect, and others, have been put forward as being of the same nature as the AB effect itself. One way or another, these claims of similarity have been shown to be doubtful or else to be misleading; various of them have been criticised.

While not devoting too much space to these topics, it seems only reasonable to give brief accounts of the proposals, and of some criticisms of them.

7.1 The Electric AB Effect

An electron wave packet passes through a diffraction slit and is split into two coherent wave packets which pass inside two long hollow metal cylinders as in Fig. 7.1. These cylinders are kept at potentials V_1 and V_2 respectively, but when the wave packets are well inside the cylinders the potentials are changed.

In particular it can be arranged that $V_1 = V_2$, except in an interval of the time τ when the wave packets are near the centre of the cylinders, during which the potential difference is altered to give $\Delta V = V_2 - V_1$. On emerging from the cylinders the wave packets come together to produce an interference pattern on the screen S.

It is claimed that in this ideal experiment the electrons have not been subject to any force due to an electromagnetic field. Also the wave packets are assumed to pass along the middle of the cylinders, well away from the walls.

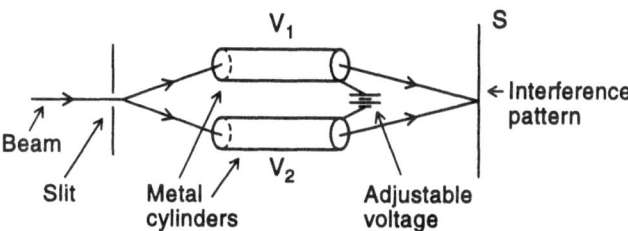

Fig. 7.1. The electric AB effect

The potential difference ΔV acting for time τ will cause the interference pattern on S to shift, due to the extra phase factor

$$\exp(-ie\Delta V\tau/\hbar) \ . \tag{7.1}$$

This is analogous to the phase factor of (2.7a) in the case of the magnetic AB effect.

The experiment was proposed by Aharonov and Bohm in Ref. [7.1]. It is pointed out in [7.2] (on pp. 27 and 120) that the experiment has not been carried out successfully. See also [7.3].

It may be that the difficulties merely arise from the nature of electron interferometry. In this connection, and concerning the question of sensitivity, there is the interesting phenomenon shown by Mateucci and Pozzi [7.4]. An electron beam is diffracted around a fine straight cylindrical filament which is coated half-and-half (lengthwise) with two different metals. Interference fringes are observed, and they move as the filament is rotated about its axis. Presumably this is due to the difference in the contact potentials of the metals.

Comments on the EAB Effect

(i) It is sometime claimed that the EAB effect has a topological aspect, because it requires the separation of the wave packet during the experiment, but not at the beginning or end of the experiment. It is claimed that there could be no EAB effect in a simply connected region of space (cf. p. 27 of [7.2] and [7.5]).

This argument may not be relevant. In the original AB effect of Chap. 2, for given induction, the phase shift its proportional to the area enclosed by the paths. That is not so for the EAB effect where the effect is instead proportional to the time τ. It is true that the two paths in Fig. 7.1 enclose an area, but the size of the phase shift is not proportional to that area.

(ii) The description of the EAB effect given above is a mixture of quantum and classical concepts, and it may be useful to state the problem more clearly.

Consider the effect of the potential V on the energy levels of an electron moving inside either cylinder. Since the electron's charge is $(-e)$ the energy levels for two different potentials change as in Fig. 7.2 on altering the potential from V to V'.

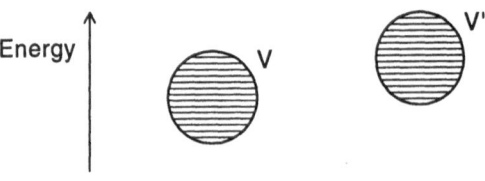

Energy

V

V'

Potentials $V' < V$

Fig. 7.2. Electron energy levels in the cylinders

The quantum form of the adiabatic theorem is stated in Sect. 8.8 below, and a detailed discussion is to be found in Refs. [8.9], [8.10]. On slowly altering a parameter of a dynamical system, the system will remain in the same quantum state, provided that this state is not degenerate. This does not mean that the state is unaltered as the parameter is changed, but it means that the state remains distinguished from neighbouring states, and it remains fully occupied.

Such an adiabatic change is indicated in Fig. 7.2 for one of the hollow metal cylinders of Fig. 7.1, in the case that the electrostatic potential is altered from V to V' (where $V' < V$). Such a change has to be made after the wave packet enters the cylinder, and it has to be reversed after time τ, before the wave packet leaves the cylinder. Only in this way can the phase change factor in (7.1) be obtained precisely.

Non-adiabatic or sudden change of the potential V will spread the system over several, or many states. In this case the process of applying the potential will not be reversible. Equation (7.1) will not be applicable.

(iii) The remarks made in (i) and (ii) show (a) that the EAB effect is likely to be difficult to detect clearly; (b) the EAB effect is of different nature from the original AB effect.

7.2 The Aharonov–Casher Effect

There is a form of symmetry between an electric charge and the magnetic moment of a neutral particle. An electric point charge moving in an induction field $\boldsymbol{B}$ is subject to a force, as in Eq. (2.1). It is simplest to think of the force acting on the charge in its rest frame.

Similarly a moving magnetic moment is subject to a force in an electric field $\boldsymbol{E}$. It is convenient to use Dirac's equation for a spin $\frac{1}{2}$ neutral particle of mass m. The Hamiltonian (using $c = 1$) is

$$H = (\boldsymbol{\alpha} \cdot \boldsymbol{p}) + \beta m \ , \tag{7.2}$$

where

$$\alpha = \begin{pmatrix} 0, & \boldsymbol{\sigma} \\ \boldsymbol{\sigma}, & 0 \end{pmatrix} \ , \qquad \beta = \begin{pmatrix} 1, & 0 \\ 0, & -1 \end{pmatrix} \tag{7.2a}$$

in the representation in which the wave function is

$$\psi = \begin{pmatrix} u_1 \\ u_2 \end{pmatrix} \ .$$

In the presence of an electric field $\boldsymbol{E}$ there will be an extra term in the Hamiltonian:

$$\mu \begin{pmatrix} 0, & -\mathrm{i}(\boldsymbol{\sigma} \cdot \boldsymbol{E}) \\ \mathrm{i}(\boldsymbol{\sigma} \cdot \boldsymbol{E}), & 0 \end{pmatrix} \ , \tag{7.3}$$

where μ is the magnetic moment, and $\boldsymbol{\sigma}$ are the Pauli spin matrices.

Dirac's equation for the electron in an electromagnetic field gives a term of the form (7.3), having the value

$$\mu = e\hbar/2mc$$

for the magnetic moment of the electron. In that case a well-known consequence of the interaction in (7.3) is the spin–orbit interaction for an electron in orbit in the electric field around a nucleus (see e. g. p. 128 of Hamilton [7.6]).

Adding the term in (7.3) to the free particle term H in (7.2) leads to the non-relativistic approximation

$$H_{\mathrm{NR}}u_1 = \frac{1}{2m}\{\boldsymbol{\sigma}\cdot(\boldsymbol{p}+\mathrm{i}\mu\boldsymbol{E})\}\{\boldsymbol{\sigma}\cdot(\boldsymbol{p}-\mathrm{i}\mu\boldsymbol{E})\}u_1 \ . \tag{7.4}$$

Simple algebra gives

$$2mH_{\mathrm{NR}} = \boldsymbol{p}^2 + \mu\boldsymbol{E}^2 - 2\mu(\boldsymbol{p}\cdot\boldsymbol{\sigma}\times\boldsymbol{E}) - \mathrm{i}\mu\hbar\boldsymbol{\alpha}\cdot\operatorname{curl}\boldsymbol{E} - \mu\hbar\boldsymbol{\nabla}\cdot\boldsymbol{E} \ . \tag{7.5}$$

The last two terms on the right of (7.5) arise from the fact that $\boldsymbol{p}$ and $\boldsymbol{E}$ do not commute in general. In a static field $\operatorname{curl}\boldsymbol{E}$ will vanish, and $\boldsymbol{\nabla}\cdot\boldsymbol{E}$ is only non-zero at the charges. For the present these terms are ignored.

The contribution of the magnetic moment to the Hamiltonian H_{NR} is

$$\mu(\boldsymbol{\sigma}\cdot\boldsymbol{v}\times\boldsymbol{E}) \ . \tag{7.5a}$$

This is the standard interaction energy for a magnetic moment moving in an electric field [7.7], [7.8]. The origin of this term is easily seen in the rest frame of the magnetic moment.

Aharonov and Casher [7.9] in effect use the non-relativistic Hamiltonian H_{NR} of (7.5) (apart from the last two terms). They consider a special system in which the axis O_z carries a uniform density Λ of electrical charge, and a magnetic dipole μ moves in any plane orthogonal to O_z. The electric intensity $\boldsymbol{E}$ has no component in the direction O_z.

The vector $\boldsymbol{p}\times\boldsymbol{E}$ is therefore in the direction O_z (or the opposite direction). Hence σ_z commutes with H_{NR}. We can choose a state in which σ_z has the eigenvalue $+1$, or the eigenvalue -1. Let this eigenvalue be s, and let $\boldsymbol{k}$ be the unit vector along O_z.

In this state we have

$$H_{\mathrm{NR}} = \frac{1}{2m}\left[(\boldsymbol{p}-\mu s\boldsymbol{k}\times\boldsymbol{E})^2 + \mu^2\boldsymbol{E}^2 - \mu^2(\boldsymbol{k}\times\boldsymbol{E})^2\right]$$

$$= \frac{1}{2m}(\boldsymbol{p}-\mu s\boldsymbol{k}\times\boldsymbol{E})^2 \qquad (s=\pm1) \ . \tag{7.6}$$

The electric field $\boldsymbol{E}$ is everywhere radially outward and of magnitude

$$\frac{\Lambda}{2\pi\rho} \tag{7.7}$$

where Λ is the linear density of charge and ρ is the radial distance from Oz. Thus

$$k \times E = E_\phi \ , \tag{7.7a}$$

where E_ϕ is a vector in the azimuthal direction (ϕ) whose magnitude is given by (7.7).

This has reduced the problem to an AB solenoid problem as in Sects. 2.1–2.3, on substituting

$$e\Phi \to \mu s\Lambda \ . \tag{7.8}$$

Thus moving the magnetic moment once around the line charge will give rise to a phase shift that can be deduced from (2.7a) by using (7.8).

Aharonov and Casher [7.9] remark that their phase shift depends on the linear charge density enclosed by the beam paths, but not on the details of the geometry of the beam paths. They state that in this sense the AC effect is topological.

A somewhat similar derivation of this symmetry has been given by Hagen in [7.10].

The use of the non-relativistic approximation, $H_{\rm NR}$ of (7.6) can be criticised in the present case since for small values of ρ the field E becomes large. The approximation in (7.4) will not be accurate in that situation, and in particular the spin dependence will not be correct for large values of E.

As it has been presented here there is an important difference between the Aharonov–Casher (AC) effect and the original AB effect of Chap. 2. In the AB effect the charged particle moves around the solenoid in a region of space where there is no Maxwell field. In the AC effect it is essential that the effective electric field E_ϕ of (7.7a) be present, and therefore the field E itself must exist at the position of the electron. For more details discussions of the AC effect see [7.11]–[7.13].

It is claimed that the AB effect vanishes in the presence of a sufficiently strong curvature of space [7.14].

7.3 The Scalar AB Effect

Strictly speaking the scalar AB effect is the electric AB effect that was discussed in Sect. 7.1 above. Since that effect is not easy to approach by experiment, a similar effect involving the interference of thermal neutrons in an induction field B has been used. A typical experiment is discussed by Allman et al. in [7.3], [7.15]. Other experiments of Aharonov–Bohm type using neutron interferometry and related topics are discussed in [7.13], [7.16]–[7.18].

The apparatus used in [7.3] for neutron interferometry is similar in principle to that shown in Fig. 7.1. The two cylinders are replaced by two solenoids, and the beam splitting, etc. is carried out by a perfect silicon crystal neutron interferometer. A collimated monochromatic neutron beam is used, with a wavelength of a few Ångstroms.

In solenoid 1 there is a constant induction whereas in solenoid 2 a periodic electric current of very short pulses gives a pulsed induction $B_2(t)$. In effect polarized neutrons are used. The potential energy of a neutron in the pulsed field is

$$V = \mu\boldsymbol{\sigma} \cdot \boldsymbol{B}_2(t) \ ,$$

and the pulses are applied while the wave packet is inside the solenoid. The expected phase shift is

$$\Delta\phi = \frac{s}{\hbar} \int \mu B_2(t) \, dt \ , \tag{7.9}$$

where the integral extends over the pulse and $s = \pm 1$ according to the neutron polarization that is used (along or opposite to $\boldsymbol{B}_2$). Also B_2 is the magnitude of $\boldsymbol{B}_2$.

It is found that (7.9) is obeyed accurately, so the SAB effect is verified precisely.

Comments

A critical discussion of this and similar experiments has been given by Peshkin and Lipkin [7.12]. They maintain that the SAB effect is merely the result of an ordinary interaction and *not* an interaction associated with an area as in the case of the AB effect. We know exactly where the neutron experienced the torque that changed the outcome of the experiment. A gauge transformation cannot alter this. They argue that the SAB effect is not a topological effect (using any useful definition of the word topological). Peshkin and Lipkin state [7.12] that somewhat similar remarks apply to the AC effect of Sect. 7.2 above. See also [7.19].

These remarks of Peshkin and Lipkin are consistent with the way Sects. 7.2 and 7.3 here have been written.

The references given in the present section should be useful to any reader wishing to know more about the remarkable neutron interference experiments.

8. Other Phenomena Involving Movement on a Closed Circuit

There is a whole variety of physical phenomena associated with the movement of a dynamical system around a closed circuit C. The circuit can be in physical space, in the space of some parameters in the problem, or in phase space [i. e. (p, q)-space]. The question arises as to whether the AB effect is similar to, or related to, some of these phenomena.

The first obvious feature is that, on account of gauge invariance, the AB effect applies only to motion around a closed curve C in space. Cyclic or periodic motions of a dynamical system have special properties that do not occur in the more general motion of the same system. In particular the action variables, and the canonical conjugate angle variables, play a basic role in cyclic or periodic motions. (It was seen in Chap. 4 that the AB effect is closely related to the properties of an action variable when magnetic induction is present.)

The classical adiabatic theorem shows that an action variable stays constant under any slow variations of the parameters of the system (with the exception of various situations that are mentioned below). The variations could be, for example, in the length of the string of a simple pendulum, or the slow rotation of the Earth in the case of Foucault's pendulum.

Anholonomy. Anholonomy is the property of a conservative dynamical system such that the situation of the system depends not only on the current values of the variables of the system, but also on the route by which the variables moved to their current values through the space of the variables concerned. That is, the situation of the system depends not only on its present state but also on its history. Anholonomy implies that the dynamical equations (of whatever sort) are non-integrable. For the case of integrable equations of motion the current values of the dynamical variables would specify the situation. Constraints of various types, or couplings of one system to another, can give rise to non-integrable equations, as we shall see. (In many works the word holonomy is used in place of anholonomy.)

Hannay's Angle. In various interesting cases in classical dynamics, on using an action variable, the classical adiabatic theroem makes it possible to show that the conjugate angle variable w does not alter by 2π on going around some closed circuit C. The discrepancy is called Hannay's angle, after the discover of this form of the effect [8.1], [8.2].

Examples of this effect are Foucault's pendulum, Larmor precession in an induction field B, and a bead running (in a plane) on a smooth wire loop which is rotated [8.1], [8.2]. Sagnac's experiment [8.3], [8.4] on the passage of light around a closed rotated mirror circuit is another example [8.5] (see [8.6] for a review of this last subject). Also the use of ring gyroscopes and ring lasers as navigation instruments is related to the present phenomena [8.7].

Time Reversal. In these examples the dynamical system is not invariant under the reversal of time. In a system without dissipation or friction, and under the effect of gravitational or electrostatic forces, the motion will run backwards in time along the same track as it ran forwards in time; it is only a matter of taking suitable initial conditions in each case. When an external angular velocity is imposed on the system, as in Foucault's pendulum, that statement is no longer true. The same holds when a fixed magnetic induction is present, because B itself reverses sign under time reversal. (B reverses sign because the current density j that is the source of B also reverses sign under time reversal.)

It is not easy to form an idea of the whole range of dynamical systems that can give rise to anholonomic effects. However we conjecture that for the *simpler mechanism*, or experiments, of this type it is necessary to violate time reversal invariance in order to have non-zero Hannay's angles, etc. (As we shall see in Sect. 8.5, anholonomy can occur in differential geometry, where time is not involved.)

Berry's Phase. A related phenomenon in wave mechanics is Berry's Phase [8.8]. This is a result of the wave mechanics adiabatic theorem [8.9], [8.10], and it applies to the slow variation of parameters around a circuit C. The wave function picks up an extra phase $\gamma(C)$. The result is closely connected to Hannay's angle [8.2]. It was pointed out by Simon [8.11] that Berry's phase can only exist for Hamiltonians that are not invariant under time reversal (or that contain imaginary terms).

Bending Light. Finally there is the fact that light, or an electromagnetic wave, that is sent around a *gently twisted* waveguide or optical fibre, may show a rotation of the linear polarisation when the ray again points in the original direction. This can be deduced by using Berry's phase, but as is explained in Sect. 8.12 below, it is a phenomenon in classical electrodynamics involving parallel transport of the polarisation triad along the waveguide or fibre.

Objectives. These various phenomena have aspects in common with the AB effect. At the end of this chapter we shall examine the extent to which the AB effect is, or is not, a consequence of any of the phenomena mentioned above. We shall also discuss a technique in wave mechanics which we call *"the phase γ"*, that is useful in the case of some anholonomic phenomena.

On account of the importance of the classical adiabatic theorem for the topics above, we must start with a brief account of the main relevant manipulations in classical dynamics theory. The chief purpose is to give some idea of the arguments for the adiabatic invariance of the action variable J, and of the methods required to make use of that property.

Readers who are familiar with action and angle variables can go directly to the Adiabatic Theorem in Sect. 8.3, or to Hannay's Angle in Sect. 8.4.

8.1 Canonical Transformations

A conservative dynamical system has the Hamiltonian $H(p_r, q_r, t)$ that is a function of the conjugate pairs of momenta and coordinates (p_r, q_r) $(r = 1, 2, \ldots n)$. H can also be an explicit function of the time t. The $2n$ equations of motion are

$$\dot{q}_r = \frac{\partial H}{\partial p_r} \quad , \qquad \dot{p}_r = -\frac{\partial H}{\partial q_r} \quad , \qquad (r = 1, 2, \ldots n) \quad . \tag{8.1}$$

In connection with Sects. 8.5, 8.10 it should be noted that the Lagrange function $L(q_r, \dot{q}_r, t)$ is related to H by

$$p_r = \frac{\partial L}{\partial \dot{q}_r} \quad , \qquad (r = 1, 2, \ldots n) \tag{8.2a}$$

$$H = \sum_{r=1}^{n} p_r \dot{q}_r - L \quad . \tag{8.2b}$$

Canonical transformations replace the pairs (p_r, q_r) by new pairs (P_r, Q_r) $(r = 1, 2, \ldots n)$, and these new pairs obey equations similar to (8.1). The purpose of such transformations is to make it easier to solve specific dynamical problems. In order to preserve the form of the equations of motion (8.1), it is necessary that

$$\sum_{r=1}^{n} p_r \, dq_r - H \, dt = \sum_{r=1}^{n} P_r \, dQ_r - K_H \, dt + dF \quad , \tag{8.3}$$

where dF is the perfect differential of any function F of the variables and the time. K_H is some function of the variables, and it can be determined when the function F is known.

The relation

$$\int_{t_1}^{t_2} dF = \int_{t_1}^{t_2} \frac{dF}{dt} dt = F(t_2) - F(t_1)$$

shows that the integral of dF only depends on the end values $F(t_2)$ and $F(t_1)$, and it does not depend on the route taken between the end points. In general the value of the integral

$$\int_{t_1}^{t_2} \frac{\partial F}{\partial t} dt$$

would depend also on the route taken between the end points.

The new equations of motion are

$$\dot{Q}_r = \frac{\partial K_H}{\partial P_r} \quad , \qquad \dot{P}_r = -\frac{\partial K_H}{\partial Q_r} \tag{8.4a}$$

and K_H is the new Hamiltonian, which is given by

$$K_H = H + \frac{\partial F}{\partial t} \quad . \tag{8.4b}$$

If F does not depend explicitly on the time, then $K_H = H$, and this quantity is the energy. Otherwise K_H will *not* be the energy of the system.

This schema can be carried out in several ways in practice. We give one example. Consider one degree of freedom ($n = 1$), and use the form $F = S(q, P, t)$. Then (8.3) can be written

$$p\,dq - H\,dt = P\,dQ - d(PQ) - K_H\,dt + \frac{\partial S}{\partial q}dq + \frac{\partial S}{\partial P}dP + \frac{\partial S}{\partial t}dt \quad . \tag{8.5}$$

In order to make the method work, PQ has been subtracted from $S(q, P, t)$ in (8.5). As $d(PQ)$ is a perfect differential, this is allowed. Equation (8.5) gives

$$p = \frac{\partial}{\partial q}S(q, P, t) \quad , \qquad Q = \frac{\partial}{\partial P}S(q, P, t)$$

$$K_H = H + \frac{\partial}{\partial t}S(q, P, t) \quad . \tag{8.6}$$

The Eqs. (8.6) describe the same system as (8.1). The problem of solving the motion has been reduced to that of determining the function $S(q, P, t)$.

8.2 Cyclic or Periodic Motion: Action and Angle

Consider a system with $n = 1$, having cyclic or periodic solutions, such that for example, orbits like those in Fig. 8.1 exist in the (p, q) plane. Rotation or libration is possible. If $H(p, q)$ is not explicitly dependent on time, any orbit is described by the equation

$$H(p, q) = W \quad , \tag{8.7}$$

where W is a constant. Varying W will give adjacent orbits, as shown in Fig. 8.1, for the cyclic case.

Equation (8.7) describes general orbits and we prefer a label that is specific to closed or periodic orbits. A suitable dynamical variable is

$$J = \frac{1}{2\pi} \oint p\,dq \quad , \tag{8.8}$$

the integral being around one closed cycle, or over one period of the motion.

For a range of orbits lying near C (Fig. 8.1), J is a function of W, and vice versa. Thus

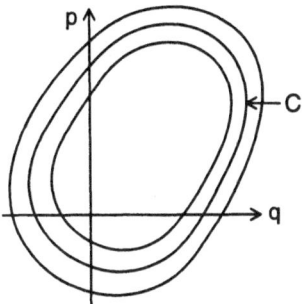

Fig. 8.1. Adjacent closed orbits in the (p, q) plane

$$J = J(W) , \qquad W = W(J) . \tag{8.8a}$$

J is constant because the orbit is closed, or the path repeats itself. We can use J as the new canonical momentum (P) and a quantity w (not to be confused with W) as the new coordinate (Q) in (8.6). Thus

$$p = \frac{\partial}{\partial q} S(q, J, t) ; \qquad w = \frac{\partial}{\partial J} S(q, J, t)$$

$$K_H = H + \frac{\partial}{\partial t} S(q, J, t) . \tag{8.9}$$

Continuing with the case in which H is not explicitly dependent on t, we can choose S to have the same property, so $K_H = H$ and (8.4a) yields

$$\dot{J} = -\frac{\partial H}{\partial w} = 0 , \qquad \dot{w} = \frac{dW}{dJ} = \text{constant} . \tag{8.10}$$

Thus it is consistent to take J as a constant of motion. w increases uniformly with time and in one orbit (or one period) w changes by

$$[w]_{(J)} = \oint \left(\frac{\partial w}{\partial q} \right)_J dq = \oint \frac{\partial^2 S(q, J)}{\partial q \partial J} dq$$

$$= \frac{\partial}{\partial J} \oint \left(\frac{\partial S}{\partial q} \right)_J dq = \frac{\partial}{\partial J} \oint p \, dq = 2\pi . \tag{8.11}$$

Equations (8.8), (8.9) have been used here. Also, $(\dots)_J$ indicates that J is constant during the differentiation, and $[X]_{(J)}$ is the change in X on passing once around the orbit $J = \text{const.}$

The canonical conjugate pair (J, w) are the angle and action variables. J labels the orbits C in phase space. For fixed time t, the angle variable w changes by 2π on moving once around the curve $J = \text{const.}$ In the motion w increases as $(\text{const}) \times t$. Equation (8.8) yields

$$J = \frac{1}{2\pi} \oint \left(\frac{\partial S}{\partial q} \right)_J dq = \frac{1}{2\pi} [S]_{(J)} . \tag{8.12}$$

For a modern mathematical treatment of this and other related topics in classical dynamics see [8.12].

An elementary example, the simple harmonic oscillator, is treated briefly in Appendix D. The particle of mass m oscillates with circular frequency ω. Then

$$S(q, J) = \int_0^q \left[2m\omega J - (m\omega q')^2\right]^{1/2} dq' \tag{8.13}$$

and (8.8) yields the form (8.8a):

$$W = J\omega \; .$$

Equation (8.9) shows how to determine $w(q, J)$ and we get the results

$$q = q_0 \sin w(q, J) \; , \qquad q_0 = \left(\frac{2J}{m\omega}\right)^{1/2} \; . \tag{8.13a}$$

By (8.10)

$$w = \omega t + \varepsilon \; , \tag{8.13b}$$

ε being an arbitrary constant. Equations (8.13a, b) give the usual equation of motion for the oscillator.

On using (8.13a), the function $S(q, J)$ in (8.13) is seen to be

$$S(q, J) = J\left(w + \tfrac{1}{2}\sin 2w\right) \; . \tag{8.14}$$

The function S therefore has a simple form when it is written in terms of w and J. It consists of a term that increases linearly with time t plus a term that is periodic in t with period π/ω. This general form of $S(q, J)$ is significant for proving the adiabatic theorem, as will be seen in Sect. 8.3 below.

8.3 Variation with Time: Adiabatic Theorem

Assume that the Hamiltonian $H(p, q, a(t))$ depends explicitly on time through the parameter $a(t)$, which can be varied as we wish. For example the circular frequency ω in a simple harmonic oscillator can be varied.

If $a(t)$ changes appreciably during a period of the system, the orbit is likely to alter considerably from those shown in Fig. 8.1. In this case there is no point in trying to define J by (8.8). However, if the change in $a(t)$ in any period is very small, the energy of the system, given by (8.7), will vary slowly; also the shape and location of the closed orbits C in Fig. 8.1 will alter slowly with time.

In such a situation (8.8) can be used to define J for each period, and we can enquire how J varies over a large number of periods. With

$$H(p, q, a(t)) = W \tag{8.15}$$

the energy W varies, and $S(q, J, a(t))$ will depend explicitly on t. Now

$$K_H = H + \frac{\partial S(q, J, a)}{\partial a}\dot{a}(t) \tag{8.15a}$$

(and K_H is *not* the energy). By (8.4a) (since H does not depend explicitly on w)

$$j = -\frac{\partial K_H}{\partial w} = -\frac{\partial^2 S(q, J, a)}{\partial w \partial q} \dot{a}(t) \quad . \tag{8.16}$$

We can write $S = S(w, J, a)$ on expressing q as a function of w and J. In Appendix E the formalism for an oscillator of more general type than the simple harmonic one, is considered.

It is shown that S is of the form

$$S(w, J) = wJ + S^+(w, J) \quad , \tag{8.17}$$

where $S^+(w, J)$ is similar to $\sin(2w)$ (8.14) in that, as a function of w, it oscillates about zero and is periodic with period π/ω.

We *assume* that this is a general property, and that it holds also for $S(w, J, a)$. By (8.16) the change ΔJ in J due to $a(t)$ varying slowly from $a(0)$ to another value $a(1)$ in time T is:

$$\Delta J = \int_0^T \dot{J}\, dt = -\int_0^T dt\, \dot{a}(t) \frac{\partial^2 S^+(w, J, a)}{\partial w \partial q} \quad . \tag{8.18}$$

Here $S^+(w, J, a)$ is periodic in w (with period π/ω) and has zero mean value.

The rate of change of $a(t)$ can be reduced on replacing $\dot{a}(t)$ by

$$\lambda^{-1}\dot{a}(t/\lambda) \quad ,$$

and increasing T to λT, where λ is a large number. The total change in $a(t)$ is unaltered, since

$$\frac{1}{\lambda} \int_0^{\lambda T} dt\, \dot{a}(t/\lambda) = a(T) - a(0)$$

$$= a(1) - a(0) \quad .$$

We assume that $S^+(w, J, a)$ can be written as a well-convergent Fourier series of the form [cf. (E.7)]:

$$\sum_n b_n \exp(2in\omega t) \quad .$$

A typical term $\exp(2i\omega t)$ contributes to the total change ΔJ, an amount

$$\frac{1}{\lambda} \int_0^{\lambda T} \dot{a}(t/\lambda) \exp(2i\omega t)\, dt = \int_0^T \dot{a}(y) \exp(2i\omega\lambda y)\, dy \quad . \tag{8.19}$$

Now λ is large, and by the Riemann–Lebesque theorem the last integral in (8.19) will vanish as $\lambda \to \infty$.

It follows that as $\lambda \to \infty$ the total change in the action ΔJ will vanish.

This is the classical dynamics adiabatic theorem. The rate of change of $a(t)$ can always be made small enough to keep J (almost) constant. The theorem does not tell us how small is "small enough". The theorem will not hold in certain cases. It fails if one of the frequencies, ω, vanishes under the variation of $a(t)$. It also fails in a multi-variable system if two frequencies ω_j and ω_k come together, or approach each other, under the variation of $a(t)$.

Several proofs of the adiabatic theorem, stating various exceptions and conditions, appear in standard texts, e. g. Fues [8.13], Arnold [8.12].

The classical adiabatic theorem is perhaps remarkable, or surprising. It is of considerable importance for periodic motion or for motion on closed loops in phase space. As a result it is important for the AB effect and for other "loop" properties such as Hannay's angle, and Berry's phase. However this statement does not in itself imply that these phenomena are closely related.

8.4 Hannay's Angle and Anholonomy

The phenomenon of Hannay's angle can be described in the following manner, although there are alternative descriptions. Suppose that the Hamiltonian of a cyclic or periodic systems depends on two parameters $\lambda(t), \mu(t)$. Starting at time $t = 0$ the parameters are varied slowly until time T when they have returned to their original values $\lambda(0), \mu(0)$ respectively. The frequency of the system $\omega(t)$ will always be (of the order of) a large multiple of T^{-1}.

The energy $W(t)$ will change slowly but the action variable J will be constant. The angle variable $w(t)$ will obey the relation (analogous to what can be deduced from the second part of (8.10) above)

$$w(T) - w(0) = \int_0^T dt \frac{\partial H(J, \lambda(t), \mu(t))}{\partial J} + \Delta\theta(J, C) \quad . \tag{8.20}$$

The integral on the right hand side just gives the effect of the (varying) circular frequency $\omega(t)$ (cf. (8.13b)). The second term on the right is Hannay's angle. It depends on the action and on the contour C described by the point $(\lambda(t), \mu(t))$ in the space of the two parameters. The fact that $\Delta\theta$ depends on C shows that the effect of these parameter changes is non-integrable.

As was mentioned in the introduction preceding Sect. 8.1, $\Delta\theta$ is only expected to exist for certain types of dynamical system. Also some kind of constraint in the dynamical system is required, although perhaps the feature concerned is not usually regarded as a constraint. The Hannay angle phenomenon will be demonstrated by various examples. References have been given at the start of this chapter.

8.4.1 Anholonomy: A Geometrical Example

From geometry, rather than dynamics, we have a simple example of this phenomenon. Consider the field of parallel vectors that are tangential to a smooth surface S (Fig. 8.2).

Consider the points P lying on a curve C that is drawn on the curved smooth surface S. At each point P there is a tangent vector that is parallel

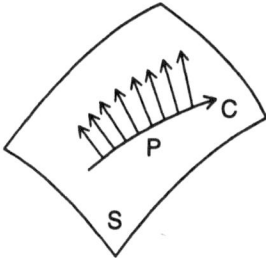

Fig. 8.2. Parallel vectors tangential to a surface S

to the tangent vectors at *immediately* neighbouring points on C. This holds for each tangential direction.

It may not be clear that this can be done. It is useful to think of the surface in the immediate neighbourhood of the point P being considerably magnified laterally. Under magnification this neighbourhood will become more like a plane, and it is then clear that parallel tangent vectors to S can be drawn at each point of C. To be more precise; if s measures distance along C, the angle between the tangents at any two points on C that are ds apart, should vanish faster than ds, as $ds \to 0$.

Although these tangent vectors are all parallel to their immediate neighbours, in general they do not lie in one plane, and the last one we draw need not be parallel to the first. If the curve C is closed, the tangent vectors on passing once around C lie (in the surface) at an angle β to the tangent vector at the start, where

$$\beta = \int_C K \, dS \ . \tag{8.21}$$

Here dS is the element of surface area, K is the Gauss curvature of the surface and the integration is over the part of S that is enclosed by C. In fact

$$K = k_1 k_2 = (\rho_1 \rho_2)^{-1} \tag{8.21a}$$

where k_1, k_2 are the principal curvatures at any point on S and ρ_1, ρ_2 are the corresponding principal radii of curvature. Thus $K = 0$ for a plane and also for any cylindrical surface. In these cases $\beta = 0$.

A textbook account of parallelism of tangent vectors with respect to a smooth surface is given in pp. 178–180 and pp. 191, 192 of [8.14]. See also, for example, [8.15], [8.16] for more recent texts. Vanishing of the (surface) intrinsic derivative, along the curve C, of the contravariant tangent vector on the surface, gives the condition for parallelism. It is easy then to show that the angle between two tangent vectors at any point P on C does not alter under parallel translation, and that the same holds for the length of the vectors. It is also easy to deduce (8.21).

A simple modern account of the differential geometry of surfaces, and of parallel vectors on a surface, is found in Chap. 12 of "Elementary Geometry" by John Roe [8.16].

For a brief mathematical discussion of parallelism see Appendix F.

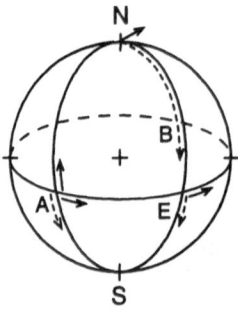

Fig. 8.3. Sphere with octant ANEA and quadrant ANBEA

A simple example of vectors tangential to the sphere is shown in Fig. 8.3. Consider the octant ANEA. Start at point A with the tangent pointing towards N. At N it will point towards B, and the same is true at E. Thus on returning to A the tangent points towards E, so $\beta = \pi/2$. As $K = R^{-2}$ and the area of the octant is $\pi R^2/2$, the same value comes from (8.21).

Another simple example is the quadrant ANBEA. The directions of the tangents are indicated by the broken arrows in Fig. 8.3. The tangent returns to A pointing towards S and $\beta = \pi$ in this case. The area enclosed is πR^2, and again (8.23) holds. On the other hand if C encloses a very small area, β will be small; this agress with the introductory statement above that in the immediate neighbourhood of a point on the surface there is no problem about parallel tangent vectors. In Sect. 8.5, around Eq. (8.27), there is an interesting mechanical example of a parallel tangential vector field on a surface.

8.5 Foucault's Pendulum and Larmor Precession

Consider Foucault's problem of the influence of the Earth's rotation on a pendulum. First use a frame of reference $Oxyz$ where Oz is the local upward vertical and Oy points South. In this frame the Newtonian equations of motion are

$$\ddot{x} - 2\alpha\dot{y} + \Omega^2 x = 0 \ ,$$
$$\ddot{y} + 2\alpha\dot{x} + \Omega^2 y = 0 \ . \tag{8.22}$$

The dots denote time derivatives in the rotating frame $Oxyz$. Also $\Omega^2 = g/l$ gives the pendulum frequency, and

$$\alpha = \omega \sin\phi \ ,$$

ω being the Earth's circular frequency while ϕ is the latitude. There are no terms of order α^2 shown in (8.22) because the Earth's centrifugal force term can be included in the gravitational terms.

On using $Z = x + iy$, the solution of (8.22) for $\alpha \ll \Omega$ is,

$$Z = ir_m \sin \Omega t \exp(-i\alpha t) \ , \tag{8.22a}$$

where r_m is the amplitude of the pendulum oscillations. This solution shows that in one period $(2\pi/\omega)$ of the Earth's rotation, the plane through Oz in which the pendulum oscillates, has rotated through the angle $(-2\pi)\sin\phi$. In the case of the North pole where $\phi = \pi/2$, the answer (-2π) is obvious (since in that case the pendulum absolutely does not rotate). For $|\phi| < \pi/2$ the situation is interesting because (a) the adiabatic theorem can be used in a characteristic fashion to obtain the angle of rotation; (b) the rotation is a simple example of anholonomy.

8.5.1 Use of the Action Variable

The Lagrangian is

$$L = \frac{m}{2}\left(v_x^2 + v_y^2\right) - \frac{m\Omega^2}{2}\left(x^2 + y^2\right) , \tag{8.22b}$$

where $(v_x v_y)$ are the absolute components of velocity v in the Oxy plane. It is convenient to write L as a function of $\dot{x}, \dot{y}; x, y$, by using the relation

$$v = \dot{r} + \omega \times r , \tag{8.23}$$

ω being the Earth's angular velocity. This gives

$$p_x = \frac{\partial L}{\partial \dot{x}} = m\dot{x} - m\alpha y ,$$

$$p_y = \frac{\partial L}{\partial \dot{y}} = m\dot{y} + m\alpha x . \tag{8.23a}$$

The action J for motion in the plane Oxy, following (8.8), is given by

$$\frac{2\pi J}{m} = \int \left(\dot{x}^2 + \dot{y}^2\right)\mathrm{d}t + \alpha\int(x\,\mathrm{d}y - y\,\mathrm{d}x) \tag{8.24}$$

$$= \int \left(\dot{x}^2 + \dot{y}^2\right)\mathrm{d}t + \alpha\int \rho^2\,\mathrm{d}\theta , \tag{8.24a}$$

where (ρ, θ) are polar coordinates in the Oxy plane. The integral in the last term in (8.24a) is twice the area in the Oxy plane that is traced out by the bob of the pendulum. Such an area term is typical of anholonomic problems.

Using (8.22a) gives the following time averages

$$\overline{\dot{x}^2 + \dot{y}^2} = r_m^2\left(\Omega^2 + \alpha^2\right)/2 , \qquad \overline{\rho^2} = r_m^2/2 . \tag{8.25}$$

The adiabatic theorem states that the action variable J, taken over any fixed number of complete periods of the pendulum, will not alter if ω is increased slowly from zero up to its physical value, provided ω/Ω stays small.

It follows that during any longish time T $(T \gg 2\pi/\Omega)$ Eqs. (8.24a), (8.25) give the relation

$$\frac{r_m^2}{2}\left(\alpha^2 T + \alpha\Theta\right) = 0 \tag{8.25a}$$

where Θ is the angle through which the plane of the pendulum rotates in T. The amplitude drops out of the calculation and hence

$$\Theta = -\alpha T \ . \tag{8.25b}$$

Taking $T = 2\pi/\omega$ (i.e. one day) gives

$$\Theta = -2\pi \sin\phi \ . \tag{8.26}$$

This is Foucault's result.

8.5.2 Anholonomy Example

This pendulum provides a simple example of the geometric anholonomy that was mentioned in Sect. 8.4 above. Equation (8.22a) can be used to give the direction of the motion of the pendulum as it passes through the origin of the coordinate system $Oxyz$. In each period of the motion, $2\pi/\Omega$, the direction of the velocity has been rotated by $-2\pi\alpha/\Omega$. One day is the multiple Ω/ω of these periods, so in that time the rotation is $-2\pi \sin\phi$.

This suggests the following consideration. Suppose that ω/Ω is so small that the surface of the Earth only moves a little during each period $2\pi/\Omega$ of the pendulum. The pendulum goes "in, out, in, out, in, ...". Only look at the "out" strokes. Suppose that these "out" strokes are marked on a fixed (imaginary) sphere surrounding the earth.

Then the "out" strokes form a field of parallel vectors tangential to this sphere. Equations (8.21) will apply, and it is easy to see that the angle of rotation of the strokes (per day) is

$$\beta = 2\pi(1 - \sin\phi) \tag{8.27}$$

as this is the area, on a sphere of unit radius, of a polar cap bounded by latitude ϕ.

Equations (8.26) and (8.27) are consistent because Θ is measured in the frame $Oxyz$, whereas β is measured on the fixed sphere. It is easy to check that each gives the correct value for $\phi = \pi/2$ (i.e. at the North pole).

Discussion. Any conservative potential $V(\boldsymbol{x})$ can occur in the Lagrangian L; however, the parameter Ω, Eq. (8.22b), does not appear explicitly in J, Eq. (8.24a), nor in the final result in (8.25b). This certainly does not mean that we could carry out the same calculation for small Ω, or for $\Omega = 0$.

Let us write $\alpha = \omega'$. We have just treated the case where $\omega' \ll \Omega$. If ω' and Ω were comparable then we could have solutions with the two periods

$$\frac{2\pi}{\left[(\Omega^2 + \omega'^2)^{1/2} \pm \omega' \right]} \ .$$

The action J cannot now be defined, as we no longer have one periodic motion that is slowly varied by another.

This prevents us from reducing Ω gradually so that we can put $\Omega = 0$ in (8.22b), and then asserting that (8.25b) is still valid, or has the same meaning

as before. We do however have the possibility of replacing the potential in
(8.22b) by some other confining potential. In that case it would be necessary
to check the new form of the averages in (8.25).

The case $\Omega = 0$ in the Newtonian equations (8.22) gives the motion
of an electron in a plane orthogonal to a uniform induction B, under the
Lorentz force of (2.1), on using $\alpha = -\omega_B/2$ (as in (8.28) below). The electron
describes a circle, but (8.25b) does not give the angular velocity. This feature
is explained in Sect. 8.10 below.

8.5.3 Larmor Precession

In Sect. 3.2 a wave mechanics solution was given for an electron moving in
a uniform induction field B. (See also Appendix H for more details of the
quantum theory solutions.) In this cyclotron motion only the Lorentz force
(2.1) acted on the electron. Here we consider another situation in which
the electron is bound to a heavy nucleus as in a two-dimensional harmonic
oscillator, in which its frequency is Ω. In addition a weak induction field
B is applied perpendicular to the plane of the oscillator. The attraction to
the heavy nucleus is the constraint on the motion of the electron under the
induction.

Using $q = -e$ for the electron in (2.1), Newton's equations of motion are
given in (8.22), but now

$$\alpha = -\omega_B/2 \ . \tag{8.28}$$

Here ω_B is the cyclotron frequency of (3.5a). The coordinate system $Oxyz$ is
now fixed in space and O is at the (fixed) nucleus. Oz is in the direction of
B. We assume that the electron moves in the plane Oxy. It is also assumed
that ω_B/Ω is small.

When a magnetic induction (along Oz) is present the action has the form
[cf. (4.7)]:

$$\frac{2\pi J}{m} = \oint \left(\dot{x}^2 + \dot{y}^2 \right) dt - \frac{e}{m} \oint \rho A_\theta \, d\theta \ . \tag{8.28a}$$

For a uniform field B along Oz we can use

$$A_\theta = \frac{\rho B}{2}$$

thus obtaining

$$\frac{2\pi J}{m} = \oint \left(\dot{x}^2 + \dot{y}^2 \right) dt - \frac{\omega_B}{2} \oint \rho^2 \, d\theta \ . \tag{8.28b}$$

This is the same form as (8.24a) on substituting for α as in (8.28).

However there is a subtle point buried here. We could use an induction
field that has the value B on a circle $\rho = \rho_T$, but does not have that value
off this circle. An example is given in Appendix AI, and below in Sect. 8.10,
e. g. (8.50). In such a case A_θ has a different form and value from that used

in (8.28b), and as a consequence the action J differs. So for the *uniform* induction B, Eq. (8.28b) gives the action; therefore we can map the problem on to Foucault's pendulum using (8.28) to give α. (For a less simple induction field this is not a valid procedure.)

Thus, when the electron moves in a plane through Oz, Eq. (8.25b) will determine the rate at which this plane rotates about Oz,namely

$$\dot{\theta} = -\alpha = +\omega_B/2 \ . \tag{8.28c}$$

If the electron has a more complicated path in the plane Oxy, this path will rotate at the rate given by (8.28c). This is Larmor's precession, which is a basic feature in atomic spectra.

In order to determine Hannay's angle Θ, the period T in (8.25b) has to be chosen as the period required for the parameters $\lambda(t)$ etc. of (8.20) to return to their initial values. This is evaluated by using the mapping on to Foucault's pendulum, namely:

$$\omega \to \omega_B \ , \qquad \phi \to -\pi/6 \ .$$

Then

$$T = \frac{2\pi}{\omega_B} \tag{8.28d}$$

and (8.25b) or (8.28c) gives

$$\Theta = +\pi \tag{8.28e}$$

and T is the time for a free electron to make one complete revolution in the plane Oxy.

The angle

$$\Theta' = -\pi \tag{8.28f}$$

is the amount by which, in time T, the electron in Larmor motion falls behind the position of the electron moving freely under the induction. Falling back by half a turn corresponds to the oscillator rotating about Oz with angular frequency $\omega_B/2$, as was given in (8.28c) above.

Comment. The same remarks concerning the parameter Ω apply here as in the section on Foucault's pendulum, just above. It is essential that ω_B/Ω (where Ω is the frequency of the electron's oscillations about the nucleus) should be small. Equation (8.28c) will not be true otherwise.

Anholonomy. This property is seen readily from (8.28a) or (8.28c).

8.6 A Bead on a Smooth Rotating Wire Loop

This is an early example of Hannay's angle [8.1] [8.2]. A bead of unit mass moves freely on a smooth wire loop. The loop lies in a plane and is rotated about an axis orthogonal to the plane with angular velocity ω (Fig. 8.4).

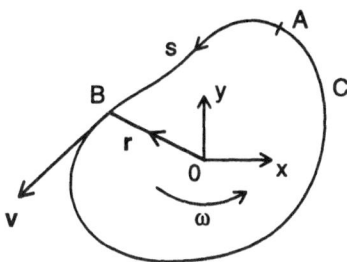

Fig. 8.4. The bead B moving on the smooth rotating wire C

The bead makes many circuits of the loop during the time in which the loop rotates once. The arc length s is measured from a datum point A that is fixed on the wire. The coordinate axes Oxy rotate with the loop, and A, L are the area and the length of the loop respectively.

The velocity v of the bead is given by (8.23) where r, $\dot{r}$ are measured in the frame Oxy. The action variable J is given by

$$2\pi J = \oint v \cdot \mathrm{d}r = \oint (\dot{r} + \omega \times r) \cdot \mathrm{d}r$$

$$= \oint \dot{r} \cdot \mathrm{d}r + 2\omega A \ . \tag{8.29}$$

The integrations are taken once around the loop.

By the adiabatic theorem J will have the same value for $\omega = 0$ as for the value of ω we choose (provided ω is small). If $\Delta\dot{s}$ is the difference in the speed of the bead between the case of $\omega \neq 0$ and the case of $\omega = 0$, we have

$$(\Delta\dot{s})_{\mathrm{AV}} = -2\frac{\omega A}{L} \tag{8.29a}$$

where the brackets denote the spatial average of the difference in speed. If $\omega > 0$, so that the sense of the circulation of the bead and the rotation of the loop are the same, as shown in Fig. 8.4, then $2\pi/\omega$ is the period of the rotation.

Using (8.29a), we estimate that by the end of this period the bead will have fallen behind the point where it would have been in the case $\omega = 0$ by the arc length

$$\Delta s = -4\pi\frac{A}{L} \ . \tag{8.30}$$

One extremum of this formula is the case of a circle of radius R for which $\Delta s = -2\pi R$. In other words, rotating the circle does not affect the motion of the bead. The other extremum is $A = 0$, for which $\Delta s = 0$. This is the case of the loop being flattened, so the bead rushes forwards and backwards on a rotating straight arm, having perfect reflections at each end.

Berry [8.2] gives an interesting discussion of the result in (8.30) by using the dynamical forces as seen in the rotating axis system Oxy (Fig. 8.4). The forces are the Corioli's, centrifugal and Euler's forces. On assuming that during any circuit of the bead around the loop, the loop itself is rotated by a very small angle, it turns out that Euler's force,

$$-m\dot{\omega} \times r$$

is the only force that contributes to Δs.

8.7 The Sagnac Effect

In the Sagnac effect a pulse or a beam of light is passed in a plane around a system of mirrors that is itself rotated with angular velocity ω, as is shown in Fig. 8.5. The time taken by a pulse to return to its starting position D (fixed on a particular mirror) is greater than it would be if $\omega = 0$ by

$$\Delta T = 2\frac{\omega A}{c^2} \tag{8.31}$$

where A is the (plane) area enclosed by the light beam.

Fig. 8.5. Sagnac effect: A light beam passes around a rotated system of mirrors

The effect was detected by sending light beams both ways around the circuit from D (Fig. 8.5). This doubled the effect, which was shown as a shift of ΔZ interference fringes where

$$\Delta A = 4\frac{\omega A}{c\lambda_0} \quad ,$$

λ_0 being the mean wavelength of the beam. The interference shift was observed by Sagnac and others [8.3], [8.4], [8.17]. Also Michelson and Gale used this device to detect the rotation of the earth [8.18]. A similar effect has been detected for neutrons [8.19].

The theory of the effect in (8.31) (Ref. [8.20]) has a general relativity form, and also what is sometimes incorrectly called a non-relativistic form. The general relativity form is found in the text books, e. g. [8.5]. A metric is chosen to describe a uniformly rotating system (that will be valid so long as the peripheral velocity stays much smaller than c). Clocks in a rotating body can only be synchronised along an open curve. If this curve is closed it will be found that, to the lowest order in ω, there is a difference (in time) between the initial and the final time of amount ΔT that is given by (8.31). The light pulse circulating around the track in Fig. 8.5 is in effect a clock; hence the result of the Sagnac experiment. The relation of this argument to anholonomy is obvious.

The simpler derivation of (8.31) is essentially kinematic. It is based on the relativistic assumption that an observer standing outside the apparatus will see the light pulse always travelling with the speed c. This is quite different from the motion of the bead on the rotating wire (above), where the speed $\dot{s}$ varied due to the rotation ω, and the difference makes the Sagnac effect simple. Let a be the length of a side of the regular hexagon in Fig. 8.5. The pulse traverses this side in time a/c for the case $\omega = 0$. For $\omega \neq 0$ the pulse has to traverse an extra length (to the lowest order in ω),

$$\delta a = \omega \frac{ab}{c} = \omega \frac{A}{3c} \tag{8.31a}$$

where $b = a(3^{1/2}/2)$ is the shortest distance from the centre O to any side. Equation (8.31) follows for the hexagonal path. Presumably a similar calculation can be made for any path, or area.

This derivation ignores all questions of possible alteration of the positions where the beam strikes the mirrors, and various other objections to the simple argument. Dresden and Yang [8.21] derive the interference formula for the case of light, and for neutrons, by studying the changes in the propagation vector due to the Doppler effect of a moving source, and to effect of moving reflectors. For a survey of the Sagnac effect, see the review by Post [8.6]. Other relevant discussions are given in [8.22]. For an application to gravitation see [8.23].

Equation (8.31) is essential for determining the sensitivity of various types of ring gyroscopes, i. e. devices for measuring rotation relative to any inertial frame [8.7]. The laser gyro is a common example [8.19]. The use of the Sagnac effect in an atomic beam is discussed in [8.24].

It should be noticed that the Josephson effect in a torus of quantum superfluid [8.25] can also be used to measure precisely small rotations, such as that of the Earth [8.26], [8.27], [8.61]. For some general practical questions relevant to precision rotation measurements see [8.28].

8.7.1 Relation of the Sagnac Effect to Other Phenomena

In the Sagnac effect, in the case $\omega = 0$, the light makes a circuit of the apparatus in the time L/c where L is the length of the light path from D back to D. This is a very short time compared with $2\pi/\omega$ the period of rotation of the apparatus. This is the situation in which we have seen Hannay's phenomenon in the examples given above.

Using ΔT of (8.31) it can be seen that

$$\frac{\Delta T}{L/c} < \frac{\omega R}{c} \quad , \tag{8.32}$$

where R is a length that is of the order of the radius of the light path. The right hand side of (8.32) is of the order of the ratio of the peripheral velocity of the apparatus to c. Thus ΔT is a very small fraction of the time L/c.

Each time the light goes around, a further delay ΔT accumulates. Thus in a period of rotation $2\pi/\omega$, the light pulse will have fallen back by a distance

$$\Delta s = -\frac{2\pi c^2 \Delta T}{\omega L} = -\frac{4\pi A}{L} \quad , \tag{8.33}$$

compared with the distance it would have gone in the case $\omega = 0$. (We can think of s as being measured from a fixed mark on a mirror.)

The maximum value of A/L (for given L) occurs for any circle of radius R, and is $R/2$. Now (8.33) becomes

$$\Delta s = -2\pi R \quad . \tag{8.33a}$$

In this case, and with the direction of rotation as shown in Fig. 8.5, the pulse of light behaves as if there were no rotation. The circular track has rotated by a distance $+2\pi R$, but by (8.33a) the light pulse has fallen back (relative to the position on the track in the case $\omega = 0$) by the same amount. This result is in agreement with the basic assumption of the kinematic derivation as given above; namely that any outside observer sees the pulse moving with speed c. This does not prevent one from seeing a fringe shift ΔZ, since that is observed at a point D fixed relative to the track.

At the opposite extreme is an apparatus for which the area A is zero, for example, if in Fig. 8.5 only an opposite pair of mirrors is used. Each mirror is bent slightly so that the light beam passes backwards and forwards between them, through the centre point O (the adjustment of the mirrors could even be varied as different values of ω are selected, if that were necessary). Clearly the light is not delayed, and the pulse does not fall back in this case. Thus Δs is zero.

Equation (8.33) is identical with (8.30) for the case of the bead on the rotating smooth wire loop; also (8.33a) applies to a circular loop. These two very different dynamical systems have the same type of anholonomic behaviour.

8.8 Berry's Phase

The discovery of this wave mechanical feature by M. N. Berry [8.2] was based on the quantum mechanical form of the adiabatic theorem [8.9], [8.10]. The theorem applies to a wave mechanical system that depends on several parameters which can be varied slowly. If the system has discrete energy eigenvalues, and if no degeneracy occurs or is approached during this variation, then the system, on being started in an eigenstate, will remain in that eigenstate. This will be true in spite of this eigenstate and its energy eigenvalue varying under the change of the parameters. Of course the theorem depends upon the rate of change in the parameters being sufficiently small (but the change itself need not be small).

It should be emphasised that this is an unusual scenario in quantum theory. The general result, on altering the parameters of a dynamical system that is in a single state, is that the system makes transitions to a set of states (with various probabilities that can be determined).

Let $R(t)$ be the parameter values at time t. It is assumed that initially the system is in the eigenstate

$$|n(R(t=0))\rangle \ . \tag{8.34}$$

The adiabatic theorem states that the system subsequently remains in the eigenstate

$$|n(R(t))\rangle \ . \tag{8.34a}$$

Notice that $|n(R(t=0))\rangle$ is a fixed state vector, and is thus a Heisenberg state vector; it merely specifies the state in which the system lies. In (8.34a), t merely specifies the value of R.

Since the system has the unusual property of remaining in a single eigenstate as time goes on, at time t it will be described by a wave function $|\psi\rangle$ which is $|n(R(t))\rangle$ multiplied by a phase factor. Berry writes this phase factor as

$$\exp[i\gamma_n(t)] \exp\left[-\frac{i}{\hbar}\int_0^t dt'\, E_n(R(t'))\right] \ . \tag{8.35}$$

The second term in (8.35) is the usual Schrödinger time factor corresponding to the instantaneous energy eigenvalue $E_n(R(t'))$, $(0 \le t' \le t)$. The first factor

$$\exp[i\gamma_n(t)] \tag{8.35a}$$

arises solely from the rate of change of the parameters, namely $\dot{R}(t)$. If R is constant this factor is unity.

By applying Schrödinger's equation to the wave function $|\psi\rangle$, Berry gets a result that can be written in the differential form

$$\delta\gamma_n = i\langle n(R) | \delta n(R)\rangle \ , \tag{8.36}$$

where $|\delta n(R)\rangle$ is the change in the basic state of (8.34) on altering R to $R + \delta R$. The normal property of

$$|n(R + \delta R)\rangle = |n(R)\rangle + |\delta n(R)\rangle$$

ensures that $\delta\gamma_n$ is a real number.

It may be necessary, or convenient, to vary R around a closed curve C in parameter space in order to detect the phase γ_n. The whole phase factor arising from (8.35a) will be

$$\exp[i\gamma_n(C)] \quad , \tag{8.37}$$

with

$$\gamma_n(C) = i \oint \langle n(\boldsymbol{R}) \,|\, \delta n(\boldsymbol{R}) \rangle \tag{8.37a}$$

in an obvious notation. The phase $\gamma_n(C)$ is sometimes called the geometric phase.

As was mentioned in the introduction in Chap. 8 (cf. [8.11]), $\gamma_n(C)$ vanishes unless time-reversal invariance is violated, or the Hamiltonian otherwise has an imaginary part. The phenomenon therefore will be anholonomic, since we can go around C in either direction.

Simon [8.11] discussed the mathematical background to Berry's phase. He points out that the adiabatic change in the parameters of the Hamiltonian produces a geometric connection in the wave functions in Hilbert space. Thus a vector $\beta(s)$ moves along the curve $C(s)$ so that the scalar product obeys

$$\langle \beta(s + \delta s), \beta(s) \rangle = 1 + O\left((\delta s)^3\right)$$

[or more precisely $(1 + o(\delta s)^2)$]. As was discussed in Sect. 8.4, this condition gives parallel transport along $C(s)$. [The parameter must be locally proportional to length along $C(s)$.]

Simon further emphasises that anholonomy arises naturally with such a connection, and the phase $\gamma(C)$ of (8.37) is what is called 'the integral of curvature', cf. (8.21).

8.8.1 Relation to Hannay's Angle

If Berry's phase and Hannay's angle exist for the same, or similar, systems, there will be a relation between them. On comparing (8.20) with the phase factors in (8.35), and labelling neighbouring states $|n(\boldsymbol{R})\rangle$ by successive integers n, the correspondence principle relation [8.2] is

$$\Delta\theta = -\partial\gamma_n/\partial n \quad . \tag{8.38}$$

This follows on noting that by the correspondence principle the unit of action is $\delta J = \hbar$.

Further progress in understanding Berry's phase requires the use of a practical and realistic example, (as in Sect. 8.9 below). For background material relevant to Berry's phase see [8.29]. Various applications to molecular physics are given in [8.30] and [8.31]. There is a survey in [8.32].

8.9 Larmor Precession and Wave Mechanics

In Sect. 8.5 Hannay's angle for Larmor precession was derived by classical methods. Here it is derived by using Schrödinger's equation.

An electron of mass m is in a two-dimensional harmonic oscillator in the Oxy plane, having circular frequency Ω. There is also a uniform induction B in the direction Oz. Schrödinger's equation is

$$i\hbar\frac{\partial\psi}{\partial t} = \left[-\frac{\hbar^2}{2m}\left(\frac{\partial^2}{\partial z^2} + \rho^{-1}\frac{\partial}{\partial\rho}\left(\rho\frac{\partial}{\partial\rho}\right)\right)\right.$$
$$\left. +\frac{1}{2m}\left(\frac{\hbar\partial}{i\rho\partial\theta} + \frac{e\rho B}{2}\right)^2 + \frac{m(\Omega\rho)^2}{2}\right]\psi \quad . \tag{8.39}$$

Here cylindrical coordinates (z,ρ,θ) are used. It will be assumed that there is no motion along Oz. The solutions of (8.39) are discussed in further detail in Appendix H (Sect. H.2), and the relation of those solutions to motion in the induction B alone is given in Sect. H.3.

For the case $B = 0$ the eigenvalue equation is separable into x and y factors. The eigenstate wave functions are

$$\psi_n(x)\psi_{n'}(y) \tag{8.39a}$$

where

$$\psi_n(x) = N_n \exp(-x^2/2a^2)H_n(x/a) \quad ; \tag{8.39b}$$

with

$$a^2 = \hbar/m\Omega \quad . \tag{8.39c}$$

N_n is a normalizing constant and H_n is a Hermite polynomial; also $n = 0, 1, 2, \ldots$. $\psi_{n'}(y)$ has a form analogous to (8.39b).

The energy eigenvalues are

$$(n + n' + 1)\hbar\Omega \quad ,$$

so in general there is considerable degeneracy in these wave functions. (See Sect. H.2 for further details on the degenerate wave functions.)

The eigenstates of (8.39) when $B = 0$, can also be found in terms of (ρ,θ) variables. For zero angular momentum (about Oz) the eigenstate is of the form

$$F(\rho^2) = \exp(-\rho^2/2a^2)L_n(\rho^2/a^2) \quad , \tag{8.40}$$

L_n being a Laguerre polynomial. For angular momentum $l\hbar$ about Oz, L_n is replaced in (8.40) by

$$\rho^l L_n^{(l)}(\rho^2/a^2)e^{\pm il\theta} \quad , \qquad (l > 0) \tag{8.40a}$$

where $L_n^{(l)}$ is a generalized Laguerre polynomial [8.33], [8.34].

8.9.1 Motion in a Weak Induction Field B

In order to find out what happens when a weak induction B is present, it is convenient to use eigenstates of the type in (8.39a). Also we should notice that anholonomic behaviour is expected to be associated with the term

$$\frac{eB}{2m}L_\theta\psi = \frac{1}{2}\omega_B L_\theta\psi \tag{8.41}$$

on the right of (8.39). This is because in order to have the particle move backwards along the same track we must reverse the sign of B. This term, (8.41), is to be equated to an extra term in $\partial\psi/\partial t$ on the left of (8.39).

A suitable eigenstate to use is

$$\begin{aligned}\Psi &\equiv \psi_n(x)\psi_0(y)\\ &= N_n N_0 \exp(-x^2/2a^2)\exp(-y^2/2a^2)H_n(x)H_0(y) \quad,\end{aligned} \tag{8.42}$$

where $H_0 = 1$. In state $\psi_n(x)$ the expectation value of x^2 is

$$\langle x^2\rangle_{(n)} = \left(n+\tfrac{1}{2}\right)a^2 \quad,$$

where a^2 is given in (8.39c).

On taking a fairly large value of n, the eigenstate Ψ of (8.42), gives oscillations along the axis Ox, together with a much smaller transverse spread of order a along Oy.

The picture of a rotating oscillator is given on substituting

$$y \Rightarrow y - \beta t \tag{8.43}$$

in (8.42): β is a constant to be determined by (8.39). On the left-hand side of that equation there is an extra term of the form:

$$i\hbar\frac{\partial}{\partial t}e^{-y^2/2a^2} = i\hbar\left(\frac{\beta y}{a^2}\right)e^{-y^2/2a^2} \quad. \tag{8.43a}$$

For small θ we put $y = \rho\theta$, giving

$$L_\theta e^{-y^2/2a^2} = \frac{\hbar}{i}\left(\frac{-y\rho}{a^2}\right)e^{-y^2/2a^2} \quad. \tag{8.43b}$$

Using (8.41), (8.43a, b) yields

$$\beta = \tfrac{1}{2}\omega_B\rho \quad,$$

so the substitution in (8.43) is determined to be

$$y \Rightarrow \rho(\theta - \omega_B t/2) \quad. \tag{8.44}$$

Repeating this process at a succession of small angular separations, it is clear that we have used Schrödinger's equation to demonstrate the Larmor precession of the oscillator in the uniform field B. Over the period $2\pi/\omega_B$ the rotation is

$$\Theta = +\pi \quad. \tag{8.44a}$$

Equation (8.44a) gives the Hannay angle Θ (or $\Delta\theta$) for Larmor precession; the same value as was derived by classical methods in (8.28e). The present method of derivation was possible because a suitable "localised" wave function was employed, namely Ψ in (8.42). This procedure is only useful if the corresponding Hannay angle exists.

We emphasise that although a wave mechanical method has been used to find Hannay's angle in this section, the Berry phase has not been derived or used.

Discussion. The constant Ω of the binding potential in (8.39), does not appear in the final answer (8.44) or (8.44a). The situation in this respect is somewhat similar to that we discussed in the classical treatment of Foucault's pendium or Larmor precession in Sect. 8.5.

It is not possible to carry through the wave mechanics calculation with Ω small, or $\Omega = 0$. The wave function $\psi_0(y)$ in (8.42) must localise the y-coordinate of the electron to within a distance a of small, or atomic, size for our calculation to make sense. Therefore by (8.39c) Ω cannot be small or zero.

8.9.2 Does Berry's Phase Exist for Larmor Precession?

We now try to derive Berry's phase for Larmor precession. A direct method will be used.

In Schrödinger's equation (8.39) the condition

$$\omega_B \ll \Omega \tag{8.45}$$

implies that the term containing $(\rho B)^2$ is everywhere small compared with the oscillator term containing $(\rho \Omega)^2$. Under the condition (8.45) the Hamiltonian is effectively linear in the induction B. The relevant term is that in (8.41).

We can proceed by looking at the change $\delta \psi$ in the wave function ψ caused by that term in any small time δt. It is

$$\delta \psi = \frac{\delta t}{2\mathrm{i}\hbar} \omega_B L_\theta \psi \; . \tag{8.45a}$$

Notice that the form of the change $\delta \psi$ is such that the normalisation of the wave function $(\psi + \delta \psi)$ is preserved. (That is because the operator in (8.41) is Hermitian.)

The Berry phase factor in (8.35a) is associated with a change $\delta \gamma$ in γ where

$$\mathrm{i}\delta \gamma \psi = \delta \psi \; .$$

Hence

$$\delta \gamma = -\mathrm{i}\langle \psi \mid \delta \psi \rangle \tag{8.45b}$$
$$= -\delta t \omega_B \langle \psi | L_\theta | \psi \rangle / 2\hbar \; . \tag{8.45c}$$

There is a reversal of sign in (8.45b) compared with (8.36). It was remarked in Sect. 8.8 that $|n(R(t = 0))\rangle$ is a Heisenberg state vector. It changes by $|\delta n(R)\rangle$ because R is altered. This change will have the opposite sign to the change $\delta \psi$ in Schrödinger's wave function (e. g. see Sect. 32 of [8.35]).

On using solutions of (8.39) having the form

$$\psi = \chi(\rho, t) \exp(iM\theta) \qquad (8.46)$$

with M an integer, Eq. (8.45c) gives

$$\delta\gamma = -\tfrac{1}{2}\delta t \omega_B M \ . \qquad (8.46a)$$

This can be integrated over the time $T = 2\pi/\omega_B$ (as was done in (8.28d) in the classical calculation). Thus we have the phase change

$$\gamma = -\pi M \qquad (M = \text{integer}) \ . \qquad (8.47)$$

Assuming for the moment that this is a Berry's phase (which it is), then by (8.38) it would correspond to the Hannay angle

$$\Delta\theta = +\pi \ . \qquad (8.47a)$$

This agrees with (8.44a) (with $\Delta\theta \equiv \Theta$). The important point is that the relation between Θ and T is linear, as in (8.25b), with $\alpha = -\tfrac{1}{2}\omega_B$ (cf. (8.28c)).

8.9.3 On the Nature of Berry's Phase

We have to consider the character of Berry's phase. It is derived for a system that has two frequencies (or cycles), a primary frequency (or cycle) Ω, that is fast, and a slow secondary frequency (or cycle) ω. The system is moved around the secondary circuit (or cycle) in an adiabatic fashion. In our derivation of (8.47), the system has to stay in the single state of the type (8.46) during time T, and that is the quantum form of the adiabatic condition.

It is not obvious that a large change in ψ can be built up from changes $\delta\psi$ as in (8.45a), without otherwise affecting the solution of (8.39). What actually happens in simple cases can be seen in connection with the phase γ that is discussed in Sect. 8.10.

Our analysis above dealt with a solution of (8.39) that is of the form

$$\psi = \exp(-iM\omega_B t/2)\psi_{\Omega M} \qquad (8.47b)$$

where $\psi_{\Omega M}$ is a wave function defined in (H.16) of Appendix H. $\psi_{\Omega M}$ is of the form shown in (8.40a) (with $M = \pm l$), and it can be written as in (8.46) above. $\psi_{\Omega M}$ describes the motion of an electron in a two-dimensional oscillator in the Oxy plane (and without any induction present). Ω is the oscillator frequency and $M\hbar$ is the angular momentum about Oz. The exponential factor on the right of (8.47b) arises from (8.46a).

When (8.45) holds, the wave function ψ of (8.47b) describes Berry's circuit in parameter space in $0 < t < T$, and the angle γ in (8.47) is a Berry phase.

This should not be confused with the phase γ that is defined in a single cycle problem in Sect. 8.10. That applies to the motion of an electron *solely* in an induction field (i. e. the case $\Omega = 0$). This is the other extreme from Larmor's problem. The Hamiltonian is divided into a part that is linear in B (and therefore gives anholonomic behaviour), and the rest – called H_0 [Eq. (8.58)] – which is unaltered by time reversal. In this case γ exists but it is not a Berry phase.

8.10 Hannay Angles, Induction Fields and the Phase γ

We first discuss the use of the Hannay angle in a field (3.20) Sect. 3.6 that is more general than the uniform induction in Sect. 8.5.3.

The use in classical physics of the action variable J_θ in closed orbits under the influence of induction has been discussed in Chap. 4. This included the relation of the original AB effect to the adiabatic principle.

The example we now give is constructed to illustrate the principles involved in applying Hannay's angle; it is not developed for practical use. The example is meant to be a partial generalisation of Larmor's model to the case of non-uniform induction.

The Lagrangian for non-relativistic motion of an electron in an induction field is [similarly to (4.5)]

$$L = \tfrac{1}{2}mv^2 - e(v \cdot A) - V(x) \ , \tag{8.48}$$

where $V(x)$ is a conservative potential. For example $V(x)$ can describe a restoring force; a possible type is

$$m(\Omega\rho)^2/2$$

in the Hamiltonian in (8.39). Or else the purpose of $V(x)$ is to guide the electron along a specified path, if guidance is required.

The induction field is of the axially symmetric type given in (4.1) and (4.1a), with

$$B_z = \frac{1}{\rho}\frac{\mathrm{d}}{\mathrm{d}\rho}[\rho A_\theta(\rho)] \ . \tag{8.49}$$

We shall be interested in particular in fields of the form (3.20)

$$B_z = \left(\frac{\rho}{\rho_{\mathrm{T}}}\right)^{n-1} B \ , \tag{8.50}$$

where B, ρ_{T}, and n are constants with $n \geq 0$.

The action variable we shall use is [cf. (4.7)]

$$J_\theta = \frac{1}{2\pi} \oint \left(m\rho^2\dot{\theta} - e\rho A_\theta\right) \mathrm{d}\theta \ . \tag{8.51}$$

As was explained in Sect. 8.5, the potential $V(x)$ does not appear explicitly in any action variable. J_θ will obey the adiabatic theorem.

There are various ways in which $V(x)$ can be chosen so that the electron can move freely on a circle $\rho = $ const. when $B = 0$.

8.10.1 Electron on an Orbit $\rho = \rho_{\mathrm{T}}$

The induction in (8.50) is now allowed to act on the electron, which is – if necessary – kept on the circle $\rho = \rho_{\mathrm{T}}$ by a (further) confining potential $V(\rho)$. This very artificial system provides a useful example. The vector potential has the form (A.2) in Appendix AI:

$$A_\theta(\rho) = \frac{B}{n+1} \frac{\rho^n}{\rho_T^{n-1}} \ . \tag{8.52}$$

The adiabatic theorem applied to J_θ shows that if B is increased slowly by δB, the value of $\dot\theta$ is increased by

$$\delta\dot\theta = \frac{e\delta B}{m(n+1)} \ . \tag{8.53}$$

Altering the induction on $\rho = \rho_T$ slowly from zero up to the value B will change the angular velocity by:

$$[\dot\theta] = \frac{\omega_B}{n+1} \tag{8.54}$$

where ω_B is the cyclotron frequency of (3.5a).

Assuming that a condition like (8.45) applies to the system, the arguments given above will yield the *Hannay angle*

$$\Delta\theta = \frac{2\pi}{n+1} \ . \tag{8.54a}$$

[Equation (8.28e) or (8.47a) is the special case $n = 1$.]

This completes our further example of a Hannay angle. The most important feature in (8.54a) is that the Hannay angle depends not only on the induction on the electron's orbit, but also on the flux through the orbit.

Starting from $\dot\theta = 0$, the change in (8.54) corresponds to a change in angular momentum of

$$\frac{1}{n+1} eB\rho_T^2 = \frac{e}{2\pi} \oint_{(\rho_T)} \boldsymbol{A} \cdot d\boldsymbol{s} \ , \tag{8.55}$$

where the integration is over one circuit around $\rho = \rho_T$. The form of (8.55) shows that the result is gauge invariant.

8.10.2 Relation of Hannay's Angle to the AB Effect

Consider the generalised AB effect in which the electron moves under the influence (only) of the induction. The term on the right of (8.55), when multiplied by $(2\pi/\hbar)$, gives the unquantized part of the total phase change $2\pi\Lambda/\hbar$ of the effective wave function $\psi_0(\theta)$ in one circuit, as is seen in (3.21) to (3.21b) in Sect. 3.7. [$\psi_0(\theta)$ is defined in (3.18a, b).] The corresponding quantized part is $2\pi M'$, where $M'\hbar$ is given by the minimum condition in (A.4), Appendix AI. This decomposition of the phase change

$$2\pi\Lambda/\hbar$$

was discussed in detail in Sects. 3.7 and 4.5.

The case of the original AB effect is essentially simpler than that above because of the fact that ρA_θ is a constant [cf. Eq. (2.3)]. A simple argument for the use of the adiabatic principle in that case has already been given in Sect. 4.5 and it yields a result of more general validity than that above. The phase γ also provides a simple method in the more general case, as will be seen below.

8.10.3 The Phase γ and Motion Under Induction

The phase γ was introduced in Sect. 8.9 in connection with Larmor precession [cf. Eq. (8.46a)]. The phase γ was directly related to the term in the Hamiltonian that violated time-reversal invariance [as per (8.45a)].

There is a similar phase γ associated with the motion of an electron in a given induction field. We shall for simplicity consider motion in the fields defined by (8.52) above [or by (A.2), Appendix AI]. If it is necessary, the electron can be guided along its path by a conservative potential $V(\boldsymbol{x})$.

We consider here motion around a single cycle. In accord with the discussion in Sect. 8.9 [between (8.47a) and (8.47b)], such phenomena do not give rise to a Berry phase.

On separating out the term on the right of (8.39) that violates time-reversal invariance, the following definition results:

$$\delta\psi = \frac{\delta t}{i\hbar} \frac{e}{m} \frac{A_\theta}{\rho} L_\theta \psi \tag{8.56}$$

$$= \frac{\delta t}{i\hbar} \frac{\omega_B}{n+1} \left\{ \frac{\rho}{\rho_T} \right\}^{n-1} L_\theta \psi \ . \tag{8.56a}$$

In order to integrate this relation we shall transform the wave function. Take the form

$$\psi = \chi(\rho, t) \exp(iM\theta) \exp(-iEt/\hbar) \tag{8.57}$$

as in (H.11a) of Appendix H. The quantity $M\hbar$ (with M an integer) is the eigenvalue of angular momentum, and E is the energy eigenvalue defined in (8.60) below.

Then

$$E\chi(\rho, t) + i\hbar \frac{\partial \chi(\rho, t)}{\partial t} = \frac{M\hbar\omega_B}{n+1} \chi(\rho, t) + H_0 \chi(\rho, t) \ , \tag{8.57a}$$

where the new Hamiltonian is

$$H_0 = -\frac{\hbar^2}{2m} \left[\frac{\partial^2}{\partial z^2} + \rho^{-1} \frac{\partial}{\partial \rho} \left(\rho \frac{\partial}{\partial \rho} \right) \right] + \frac{1}{2m} \left[\left(\frac{M\hbar}{\rho} \right)^2 + (eA_\theta)^2 \right] \ . \tag{8.58}$$

Clearly H_0 contains no term that violates time-reversal invariance.

For ρ near ρ_T we write

$$\chi(\rho, t) = \exp \left(-\frac{iM\omega_B t}{n+1} \right) \chi_0(\rho) \tag{8.59}$$

and the function $\chi_0(\rho)$ obeys

$$H_0 \chi_0(\rho) = E\chi_0(\rho) \ . \tag{8.60}$$

A term on the right of (8.57a), of the form

$$M\hbar \frac{\omega_B}{n+1} \left[\left(\frac{\rho}{\rho_T} \right)^{n-1} - 1 \right] \chi(\rho, t) \ , \tag{8.60a}$$

has been neglected here. The wave function $\chi_0(\rho)$ has to be chosen so that the expectation value formed from the term in (8.60a) is small.

The Result of the Transformation. On using the idea of a wave packet, we know that the electron completes a circuit on $\rho = \rho_T$ in the period $T = 2\pi/\omega_B$. Equation (8.59) shows that in time T, $\chi(\rho, t)$ has acquired an extra change in phase of

$$\gamma = -\frac{2\pi M}{n+1} . \tag{8.61}$$

[The sign of γ also follows from the notation in (8.47) for the case $n = 1$.] This quantity is the phase γ.

By (8.38) we can relate an angle to this phase γ. This angle will be called $\Delta\theta$, (analogous to Hannay's angle, just as γ is analogous to Berry's phase). Then

$$\Delta\theta = \frac{2\pi}{n+1} . \tag{8.61a}$$

A physical interpretation, coming from (8.54) above, is that $\Delta\theta/2\pi$ is the fraction of the total angular momentum, $\Lambda = m\omega_B\rho_T^2$ of an electron constrained to move on $\rho = \rho_T$ that can be generated by an adiabatic increase in the induction alone.

Why does (8.61a) give the correct answer for the fraction of Λ that is generated adiabatically? [It is clear that the only adiabatic feature in our derivation of γ is that the wave function ψ of (8.57) remains in one eigenstate of L_θ.] The answer to this puzzle is that in deriving γ here, we are in fact finding the AB phase angle (in the generalised AB case). In connection with the discussion following (8.55) above, the AB phase angle is closely related to the fraction of Λ that can be generated by slowly increasing B.

A Check. A rough but independent method is used in Sect. 8.11.1 below to show that the phase-γ method is sensible.

Example (i) The Phase γ and the Generalised AB Phase (The Case $\rho = \rho_T$)

We have used the wave function

$$\psi = \exp\left(-\frac{iM\omega_B t}{n+1}\right) \chi_0(\rho)\,\exp(iM\theta)\,\exp(-iEt/\hbar) \tag{8.62}$$

as taken from (8.57), (8.59). This differs from the wave function used in Chap. 3 or Appendix H, in that in (8.62) E is the eigenvalue of the time symmetric Hamiltonian H_0, as in (8.60).

The Hamiltonian H_0 contains no term linear in L_z, therefore $M\hbar$ itself has to be the eigenvalue of *angular momentum* for the wave function in (8.62).

For motion on, or close to, $\rho = \rho_T$, this gives

$$M\hbar = m\omega_B\rho_T^2 . \tag{8.63}$$

Notice that quantisation here differs from that in (3.4), and from that in (3.20b) (or (A.4), Appendix AI).

It follows explicitly from the form of A_θ in (8.52) that

$$\frac{e}{2\pi} \oint_{(\rho_T)} \boldsymbol{A} \cdot d\boldsymbol{s} = \frac{M\hbar}{n+1} \quad . \tag{8.64}$$

Thus by (8.61)

$$\gamma = -\frac{e}{\hbar} \oint_{(\rho_T)} \boldsymbol{A} \cdot d\boldsymbol{s} \quad . \tag{8.65}$$

Hence the phase factor $\exp(i\gamma)$ is just the AB phase factor of (2.7a), as applied to the generalized case of motion of an electron in the field given by (8.50).

Example (ii)   The Original AB Effect and the Phase γ

Consider an electron *wave packet* moving in the Oxy plane on a closed orbit; it passes once around the cylindrical solenoid whose axis is along Oz. The potential $V(\rho)$, which is independent of θ, guides the electron, but in general the orbit will not be a circle.

The situation is shown in Fig. (8.6). The coordinate (z, ρ, θ) will be used. The method of (8.56) picks out the anholonomic part of the change in the wave function ψ during the small interval of time δt. This is

$$\delta\psi = \frac{\delta t}{i\hbar} \frac{e}{m} \frac{\Phi}{2\pi\rho^2} L_\theta \psi \quad , \tag{8.66}$$

where (2.3) has been used for A_θ, and Φ is the flux in the solenoid.

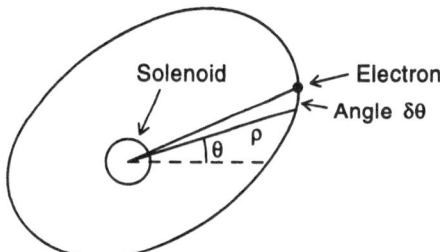

Solenoid

Electron

Angle $\delta\theta$

θ ρ

Fig. 8.6. The electron and the solenoid

Equations (8.66) and (8.45c) indicate an increment $\delta\gamma$ in the phase γ, with

$$\delta\gamma = -\frac{\delta t}{\hbar} \frac{e}{m} \frac{\Phi}{2\pi\rho^2} \langle \psi | L_\theta | \psi \rangle \quad . \tag{8.66a}$$

A wave function ψ that is a single eigenstate of L_θ will not show motion around the orbit C. It is necessary for ψ to be a superposition of eigenstates having adjacent eigenvalues $M\hbar$ of L_θ. A small spread of such eigenstates can depict motion around the solenoid in a fashion that does not violate the uncertainty principle. Let $\langle M \rangle$ be the mean value of M in this wave packet. Equation (8.66a) becomes

$$\delta\gamma = -\delta t \frac{e}{m} \frac{\Phi}{2\pi\rho^2} \langle M \rangle \quad . \tag{8.67}$$

Because $V(\rho)$ is independent of θ and there is no other force on the electron, the angular momentum about Oz obeys

$$m\rho^2\dot{\theta} = \text{const.} = \langle M \rangle \hbar \tag{8.67a}$$

on the curve C. So (8.67) can be written as

$$\delta\gamma = -\frac{e}{\hbar} \frac{\Phi}{2\pi} \delta\theta \quad . \tag{8.67b}$$

Here $\delta\theta$ is the change in the angle θ in the interval δt.

For one complete circuit of C we have

$$\gamma = -\frac{e\Phi}{\hbar} \quad . \tag{8.68}$$

This is just the original (solenoid) AB effect, as was displayed in (2.7a). [This result is not to be confused with (8.65) which has the identical form, but relates to the generalised AB effect of Chap. 3.] Our derivation requires the orbit to be closed, but it need not be a circle.

As (8.67a) is essential for this derivation, the method only applies to a circulating *wave packet*. The AB effect however holds for any wave function that surrounds the flux, as was shown in Chap. 2.

Comment. The question arises as to wether the AB phase can actually be expressed as a time-dependent term like the phase γ. It is clear from the discussion in Chap. 2 (Sects. 2.2 to 2.4) that for a simple interpretation of the AB effect, it is necessary that the confining potential is of the form $V(\rho)$ and it does not depend on θ.

For a wave packet, (8.67a) then follows, and this simply relates the t-form and the θ-form of the phase factor.

8.10.4 Principles of the Phase γ and Conclusions

In Schrödinger's equation the term linear in L_θ is the anholonomic term. Under the conditions: (i) the wave function ψ is an eigenstate of L_θ, (ii) the wave packet $\chi(\rho, t)$ can be chosen so that (the expectation value of) $\rho^{-1}A_\theta(\rho)$ varies little during a circuit $0 \leq \theta \leq 2\pi$, then the anholonomic term in the Hamiltonian can be replaced by the factor

$$\exp(-i\alpha t) \tag{8.62a}$$

in the wave function ψ, α being a constant.

Then ψ is of the same general form as the wave functions in (8.47b) and (8.62). The remaining part of the Hamiltonian has no anholonomic term, so the whole anholonomic effect is now in the factor in (8.57).

In the case of weak uniform induction (Larmor's problem) the phase γ is equal to Berry's phase.

For the motion of an electron in various induction fields the phase γ equals the AB phase or the generalised AB phase.

The calculation of the original AB effect (the solenoid problem) via the phase γ is shown in a general method in Example (ii) above.

8.11 Miscellaneous Topics

8.11.1 The Energy Estimate Used to Check the Method of Sect. 8.10

Since the method used in (8.57)–(8.60) for separating off the anholonomic term in the Hamiltonian is important, it is useful to check its practical value, by estimating the energy of the state ψ of (8.57).

The variational method of solution is discussed in Sects. 3.3 and 3.6, and Appendix AI. The wave function χ is assumed to be independent of z, and the term in H_0 containing $\partial/\partial\rho$ is assumed to relate to the oscillations in the radial direction (i. e. oscillations in ρ) as in Sect. 3.3 and Eqs. (A.6a–c) in Appendix AI.

The remaining part H_0' of H_0 is of the form

$$H_0' = \frac{1}{2m}\left\{\frac{A}{x} + Cx^n\right\} , \qquad (x \equiv \rho^2)$$

where A, C are given constants. Taking the minimum value of H_0', gives

$$\text{Min}(H_0') = \frac{mv_{\text{T}}^2}{2(n+1)} , \tag{8.69}$$

where v_{T} is the electron's velocity on $\rho = \rho_{\text{T}}$. Requiring the minimum to occur at $\rho = \rho_{\text{T}}$ yields the condition

$$M\hbar = \frac{n^{1/2}}{n+1}\Lambda \tag{8.69a}$$

where Λ is the angular momentum,

$$\Lambda = \pm eB\rho_{\text{T}}^2 ,$$

and M is the integer defined in (8.57).

The energy associated with the anholonomic term in (8.59) is

$$E_{\text{AN}} = \frac{M\hbar\omega_{\text{B}}}{n+1} . \tag{8.69b}$$

Thus

$$\text{Min}(H_0') + E_{\text{AN}} = \frac{1}{2}mv_{\text{T}}^2\left(\frac{n^{1/2}+1}{n+1}\right)^2 . \tag{8.69c}$$

The last factor on the right of (8.69c) has the values:

1.44	1.29	1.00	0.65	0.47

for $n =$ 1/4   1/2   1   2   3   .

Conclusion. For values of n between $1/2$ and 2 the approximations we have made in using (8.57), (8.59) and (8.60), seem to be reasonable. Further details are given at the end of Appendix AII.

The values of M, H_0', etc. in (8.69)–(8.69b) are only to be used for this approximation, whose sole purpose is to show that our method makes sense numerically.

For cases with $n \neq 1$, the cross-term [involving $L_\theta A_\theta(\rho)$] varies with ρ, and it makes a difference whether the minimum is taken with this term includeed (as in Chap. 3) or without it (as in H_0'). Hence the quantum number M in (8.69a) does not agree with M' in (A.4).

The Sense of Motion. The Hamiltonian H_0 in (8.58) contains B^2, but not $\boldsymbol{B}$. Since the sense in which the electron goes around the orbit $\rho = \rho_{\mathrm{T}}$ reverses on reversing the sign of $\boldsymbol{B}$, it follows that the anholonomic term in (8.59) must determine the sense of the motion.

On reversing the sign of $\boldsymbol{B}$, the value of ω_{B} will change sign. For positive B (i. e. $\boldsymbol{B}$ pointing along the axis Oz) the standard solution of Sect. 3.1 gives M positive, as is seen in (3.4). Therefore the correct sign of the exponent in (8.59) will require $M\omega_{\mathrm{B}} > 0$.

Now look at the solution we have just given. The value of $\mathrm{Min}(H_0')$ in (8.69) is always positive. Consider what happens when the sign of E_{AN} in (8.69c) is reversed. The right-hand side of that equation takes the new value

$$\frac{1}{2}mv_{\mathrm{T}}^2 \left(\frac{n^{1/2}-1}{n+1}\right)^2 . \tag{8.69d}$$

By (8.69b) this is the estimate of the energy for the case $M\omega_{\mathrm{B}} < 0$.

For $n > 0$ the quantity in (8.69d) cannot give the value $\frac{1}{2}mv_{\mathrm{T}}^2$, as is required for a circulating electron. It follows that $M\omega_{\mathrm{B}}$ has to be positive to give a physical solution here. In the present method the anholonomic term is the cause of this result.

8.11.2 Particle Accelerators

The results arising from using the action J_θ, Eq. (8.51), are related to basic properties of accelerators for charged particles on circular tracks. The charged particle is constrained to move on a circular, or nearly circular, track by a confining potential V. It is assumed that this potential is a function $V(\rho)$ of ρ alone. Here ρ is the distance from the centre O of the track, which lies in a plane. We shall assume that the particles are electrons (of charge $-e$).

Equation (8.54) gives the maximum angular velocity to the obtained in such an accelerator by increasing B slowly, and (8.55) relates the angular momentum Λ about O to the induction field. This field is given by (8.49) and (8.50) and the constant n obeys $n \geq 0$. The electron is constrained to stay on the circle $\rho = \rho_{\mathrm{T}}$. The remaining angular velocity, or angular momentum (needed to reach the appropriate value ω_{B}), must be added by electric fields, accelerating cavities or other devices. This second part of the angular momentum corresponds to the quantized terms $M\hbar$ (or $M'\hbar$) in the wave mechanics solution [cf. (A.4) or (3.20b) and (3.21b)].

By (8.61a), $\Delta\theta/2\pi$ is the *fraction* of the total angular momentum (of a particle circulating on an orbit $\rho = \rho_T$) that can be produced solely by changing B slowly. For $n = 0$ we have the betatron case that was already discussed in Sect. 3.8. As is seen in (3.22), the field in such accelerators has to obey,

$$\Phi(\rho_T) = 2\pi\rho_T^2 B$$

for motion on the orbit $\rho = \rho_T$. In this type of accelerator there is no need for cavities or other similar accelerating devices (except for compensating the energy losses).

The other extreme case is large n. Then $\Delta\theta/2\pi$ is very small. This is the situation in which B is very small inside $\rho = \rho_T$, but it has the magnitude B just on the circle. All the acceleration will have to be produced by cavities, or the like. Also there is almost no AB effect in this case.

A standard relation in the theory of circular accelerators is (cf. Sect. 2.1 of [8.36])

$$\frac{d}{dt}\left\{ m\rho^2\dot\theta - \frac{e\Phi}{2\pi} \right\} = \frac{dw}{d\theta} - \frac{dL}{d\theta} \ , \tag{8.70}$$

where Φ is the flux of induction through the orbit, $dw/d\theta$ is the torque due to the applied electric field and $dL/d\theta$ is the decelerating torque arising from the energy lost by radiation due to radial oscillations about the mean orbit, etc. Equation (8.70) has considerable similarities with our discussion above.

The left-hand side of (8.70) is related to the rate of change of J_θ (cf. (8.51)). The term containing Φ is associated with the generalised AB effect, as was discussed in Sect. 2.7 [in particular notice (2.20a)]. The right-hand side of (8.70) is related to the changes in the quantised terms $M\hbar$, $M'\hbar$, in the wave mechanics solution of Sects. 3.1, 3.6 and 3.7.

8.11.3 The Original AB Effect and Berry's Phase

The original AB effect (of Chap. 2) and Berry's phase are associated with motion around a closed circuit in some space. Also they both involve the violation of Wigner's time reversal. First we will give the reasons for believing that nevertheless the original AB effect and Berry's phase are distinct phenomena based on different principles.

It is sometimes said that the AB effect is a consequence of Berry's phase. It seems improbable that this can be true for the following reason: The AB phase does not depend on how slow or how fast the electron (or the electron wave packet) goes around the closed circuit C. Its existence does not depend on how slow, or how fast, the induction B has been built up. The AB phase does not depend on any adiabatic principle whatever, classical or quantum. The

AB phase does not require the electron's state to be set up slowly or smoothly. Basically it depends only on the simplest properties of wave mechanics and electrodynamics.

Berry's phase depends on adiabatic properties, and adiabatic change (or lack of change) is not a simple concept, in its classical or in its quantum form.

As was emphasised above, in the paragraph in Sect. 8.9 between (8.47a) and (8.47b), and elsewhere, the obvious feature is that Berry's phase is associated with two cycles, one fast and the other slow. The AB phase is associated with only one cycle (unless one goes in for unessential elaboration).

From quite another point of view, a difference between Berry's phase and the AB phase is illustrated by the following. In Ref. [8.37] deviations from Berry's adiabatic geometric phase in a ^{131}Xe nuclear gyroscope are observed and discussed. A similar paper on deviations from an AB phase, if it were correct, would be remarkable and startling, since the AB phase must be exact if the current basic ideas about wave mechanics and electromagnetism are true.

Berry [8.8] puts a particle of charge q into a coordinate box and the origin of the box is transported on a circuit C around a line of flux. This gives a phase factor of the form (8.37), (8.37a) where $\gamma(C)$ is the usual AB phase

$$q\Phi/\hbar \ ,$$

Φ being the flux through C. However this derivation itself starts from a relation that is identical with (2.6). That relation between the solutions $\psi_A(x)$ and $\psi_0(x)$ of the two Schrödinger equations is of course the starting point of Aharonov and Bohm [8.38], [8.39]. It is independent of any adiabatic ideas.

Briefly, Hannay's angle and Berry's phase are based on two adiabatic theorems. (So for example, Foucault's pendulum result would not hold for a rapidly rotating earth). The AB effect is valid whatever way the electron moves since it depends only on the basic relation (2.2) in Chap. 2.

8.12 Anholonomy and Electromagnetic Wave Polarization

The rotation of the linear polarization of light, or electromagnetic waves, on passing around a coiled light fibre or wave guide, has been given as an example of Berry's phase [8.40]. In fact (as Berry himself points out [8.29], [8.41]) this phenomenon has an older and independent history.

Rytov in 1938 [8.42] discussed the classical propagation of electromagnetic waves in inhomogeneous media in the limiting short wave case. He showed that the directions e, h of the electric and magnetic fields moved by parallel transport of the triad of vectors e, h, t, where t is the ray direction. Vladimirskii [8.53] showed that this law is not integrable, and discussed the rotation of polarisation that resulted.

If the light fibre were a closed curve C lying in a plane, then in one circuit the polarisation triad (e, h, t) will be rotated through 2π around the normal to the plane, and no effect is detectable. So the light fibre has to be an open curve C such that t will return to its original *direction*, once or several times; for example as in a helix.

Born and Wolf (cf. Sect. 3.1.3 of [8.43]) point out that in a medium having a smoothly varying refractive index n, the electromagnetic vectors are parallel transported along a light ray. [No guide or walls are used in this derivation.] Specifically they state that in non-Euclidean geometry, whose line element is

$$ds' = n\,ds = n\left(dx^2 + dy^2 + dz^2\right)^{1/2} \quad ,$$

the geometrical light rays are geodesics and the two polarisation vectors are parallel transported along each ray (cf. Bortolotti [8.44]).

The "smoothly varying" property is analogous to the requirement of adiabatic change in preceding sections. A mirror will reverse the helicity of light, and a sharp change in n will similarly cause some change in the helicity. Since rotation of linear polarization implies some difference in the phase velocities of the two circularly polarized rays, only a smooth change in n is permissable. Similarly only gently coiled fibres or guides are to be used in experiments. (See also Ref. [8.45] for the basic waveguide ideas.)

On a smooth curve C at any point P there are the tangent t, the (principal) normal n and the binormal b; (t, n, b) is a right handed orthogonal triad of unit vectors. n lies in the osculating plane at P (i. e. the plane that contains three adjacent points of C close to P), and b is orthogonal to it. For a plane curve, b is orthogonal to the plane. In general

$$\frac{db}{ds} = -\tau n \quad , \tag{8.71}$$

where s is the arc length and $\tau(s)$ is the torsion at P. Positive τ means that b rotates in a positive screw sense about t as s increases.

As P moves along C, the osculating plane will move, and it traces out the surface called the envelope. Adjacent osculating planes at P intersect in the tangent t at P; therefore the tangents to C trace out the envelope. It follows that the envelope of the osculating planes is a developable surface. It is called the osculating developable (see e. g. Ref. [8.46]). Each osculating plane touches the envelope. This picture describes the movement of the orthogonal triad (t, n, b) as P moves along C.

Under reasonable assumptions, including gentle bends in the fibre and smoothly varying refractive index, it has been shown [8.41] that the polarisation of an electromagnetic wave propagating in the fibre along the positive direction of s will rotate (relative to n and b) in the opposite sense to the rotation given by (8.71). The rate of rotation in the two cases is approximately the same. This means that the polarisation of the electromagnetic pulse describes parallel transport along C.

Allowing for the fact that, for a closed plane curve, the rotation of the polarisation triad is 2π, the angle Ω of rotation of the polarisation along a segment C_1 of the fibre is [8.41], [8.47]

$$\Omega = 2\pi - \int_{C_1} \tau(s)\,ds \ .$$ (8.72)

This segment C_1 of C is such that the tangent t makes exactly one complete revolution as P traverses C_1.

For a circular helix of radius a and slope angle α we have

$$\tau = \frac{1}{2a}\sin 2\alpha \ , \qquad L = \frac{2\pi a}{\cos\alpha} \ ,$$

L being the arc length of one turn. Equation (8.72) gives

$$\Omega = 2\pi(1 - \sin\alpha)$$ (8.73)

for the single turn. The extreme case $\alpha = 0$ is the circle, while $\alpha \to \pi/2$ is, in the limit, an infinite straight fibre that is parallel to the axis of the helix, for which $\Omega = 0$.

In the case of the circular helix, the binormal b has a circular motion as P traverses C_1. On holding the base of b fixed, the head of b describes a circle of radius $\sin\alpha$. Minus the circumference of this circle is the contribution of b to Ω. [In a similar fashion the unit vector t traces out a circle that encloses a solid angle Ω given by (8.73).] It is to be noted that (8.73) is similar to (8.27) for Foucault's pendulum.

Practical experiments that verify the rotation of optical polarisation are described in Refs. [8.48]–[8.51].

There is a further problem in relation to the experiments. Fibres that allow a single mode only are used, and ray optics (which is used directly above) may not be applicable. It is however possible to give a simple treatment that gets around the difficulty [8.41]. Further interesting and relevant work on short wavelength radio propagation can be found in [8.52].

For further discussion on anholonomy in optical problems see the article by Chiao in [8.48].

The is more detailed information on the parallel transport of vectors orthogonal to a twisted curve in the second part of Appendix F. The work of I. and Z. Bialynicki–Birula [8.54] on parallel transport and spin in the relativistic case also describes the motion of photons in twisted guides.

8.13 Further Developments

8.13.1 Generalization of Berry's Phase

There are interesting generalizations of Berry's phase in several directions. An important one is the effect of degeneracy of the energy eigenvalues of Schrödinger's equation.

The simplest example is found in Berry's original paper [8.2]. Looking apart from symmetry it is in general necessary to vary three parameters in

the Hamiltonian in order to cause two energy levels to coincide. Suppose that
this occurs at a point D in the three-dimensional parameter space. Consider
a closed curve C in this space that lies near D. Then the geometrical phase
$\gamma(C)$ of (8.37) equals

$$\pm \tfrac{1}{2}\Omega(C) \ , \tag{8.74}$$

where $\Omega(C)$ is the solid angle subtended at D by C, and the $\pm$ sign depends
on which of the two eigensolutions (or eigenstates) is used.

If the Hamiltonian is purely real on C, the three-dimensional space reduces
to a plane. Clearly $\Omega = \pm 2\pi$ if D is inside C, and $\Omega = 0$ otherwise; so the
quantity in (8.37) equals -1 or $+1$ respectively in these two cases.

A quite different generalization is due to Wilczek and Zee [8.55]. Suppose
that there are a number of parameters

$$\boldsymbol{\lambda} = (\lambda_1, \lambda_2, \lambda_3, \ldots)$$

in the Hamiltonian, and the parameters can be varied in a continuous manner.
Suppose further that for all relevant values of $\boldsymbol{\lambda}$ the Hamiltonian has a set of
n energy levels that are degenerate. This can be due, in whole or in part, to
symmetry. It is also possible to arrange that as $\boldsymbol{\lambda}$ varies the eigenvalue stays
at a value E_0.

The parameters $\boldsymbol{\lambda}$ are varied slowly from the initial value $\boldsymbol{\lambda}_I$ around a
circuit C in $\boldsymbol{\lambda}$-space, and finally back to $\boldsymbol{\lambda}_I$.

For sufficiently slow variation the quantum adiabatic theorem is applied
to the solution of Schrödinger's equation

$$i\hbar \partial \psi / \partial t = H(\boldsymbol{\lambda}(t)) \psi \ . \tag{8.75}$$

If the group of degenerate levels does not cross any other energy level,
the adiabatic process maps the original n levels on to n linear combinations
of themselves. The mapping is of the type $U(n)$. This is the generalization of
Berry's result in (8.37).

Wilczek and Zee [8.55] show that, subject to further conditions which we
shall not give here, the matrices $U(n)$ in $\boldsymbol{\lambda}$-space are equivalent to a set of
integrals around a Wilson loop, the integrands A_μ being non-Abelian gauge
potentials. For further discussion, see [8.56].

8.13.2 Other Topics

The problem of geometric phases for the relativistic theory of spinning particles was discussed by I. and Z. Bialynicki–Birula [8.54] and by Jordan [8.57].
A universal connection in momentum space determines the parallel transport
of the spin during the motion of the particle through external fields. The motion of electromagnetic radiation in a twisted guide is another example, as is
seen in Sect. 8.12 above.

Mathur [8.58] relates the spin-orbital interaction in Dirac theory as applied to an atom, to Berry's phase. The orbital motion is here assumed to

change slowly relative to the particle to anti-particle fluctuations. Shankar and Mathur [8.59] have related Thomas precession in the non-relativistic limit to a non-Abelian Berry vector potential.

In this terminology, the integral on the right of (8.37a) in Sect. 8.8 is an integral over a Berry vector potential. Also, the calculation leading to (8.74) in the case of two degenerate levels, shows that in this case the potential arises from a monopole situated at the point D.

For the general relation of such problems to time reversal invariance, see Avron et al. [8.59].

Jordan [8.57] gives a simple description of the formalism in the case of relativistic particles.

Field Theory and Differential Geometry. A simple account of non-Abelian gauge fields, their relation to differential geometry, and the connection that defines parallel displacements, is to be found in Chap. 12 of the textbook "Quantum Field Theory" by Itzykson and Zuber [8.60]. A similar account is found in Chap. 1 of "Gauge Fields" by Faddeev and Slavnov [8.62].

Appendix AI. Symmetric Motion in Axially Symmetric Fields by Quantum Extremum Method

This appendix shows how the quantum extremum method that was used in Sects. 3.1–3 for the case of motion in uniform induction, can be applied to two-dimensional motion in an axially symmetric induction in the case of symmetric orbits. The cyclotron oscillations about such orbits are also derived.

Consider an induction field of the form

$$\boldsymbol{B} = (0, 0, B_z) \ , \qquad B_z = \left(\frac{\rho}{\rho_{\mathrm{T}}}\right)^{n-1} B \ , \tag{A.1}$$

where n is a positive integer and ρ_{T} is a fixed radius. In cylindrical coordinates (z, ρ, θ) we use

$$\boldsymbol{A} = (0, 0, A_\theta) \ , \qquad A_\theta = \frac{B}{n+1} \frac{\rho^n}{\rho_{\mathrm{T}}^{n-1}} \ . \tag{A.2}$$

For circular motion about O_z we use the single-valued wave function

$$\psi = \chi(\rho, t) \exp(\mathrm{i} M' \theta) \ , \qquad (M' = \text{integer}) \ . \tag{A.3}$$

The wave equation is the same as (3.3a), except that the last term on the right is

$$(2m)^{-1} \left(\rho^{-1} L_z + e A_\theta\right)^2 \psi \ , \tag{A.3a}$$

with A_θ as in (A.2). By (A.3),

$$\left(\rho^{-1} L_z + e A_\theta\right) \psi = \left(\frac{\hbar M'}{\rho} + \frac{e B \rho^n}{(n+1)\rho_{\mathrm{T}}^{n-1}}\right) \psi \ . \tag{A.3b}$$

The factor in front of ψ on the right of (A.3b) has a single extremum for $M'/B > 0$, and none otherwise. The extremum occurs at $\rho = \rho_{\mathrm{T}}$ (where the induction is $B_z = B$), provided that

$$M' \hbar = \frac{n}{n+1} e B \rho_{\mathrm{T}}^2 \ , \tag{A.4}$$

and the extremum value of the term in brackets on the right of (A.3b) is

$$e B \rho_{\mathrm{T}} \ . \tag{A.4a}$$

Equation (A.4) gives the quantization of ρ_{T}. There is no extremum for the betatron case, $n = 0$ (cf. the discussion in Sect. 3.8).

The relation between extrema of the Hamiltonian and its eigenstates shows that, apart from vibrations in ρ, the electron moves in a positive sense about O_z for $B > 0$. It moves in the induction $B_z = B$ on the circle $\rho = \rho_T$. By (A.4a) its peripheral momentum P_θ and the kinetic energy T_θ are

$$P_\theta = eB\rho_T \; , \qquad T_\theta = (eB\rho_T)^2/2m \; . \tag{A.4b}$$

On replacing ρ_0 by ρ_T it is seen, from (3.6) and (3.8b), that these are the same values as for motion in the uniform field B. However for $n \neq 1$, T_θ not equal to $M'\hbar\omega_B$, where ω_B is the cyclotron frequency of (3.5a). This difference from motion in the uniform field is due to the generalised AB effect.

In terms of (3.18)–(3.18b) the wave function in (A.3) is $\psi_A(x)$. For movement once around the circle $\rho = \rho_T$ we have

$$\frac{e}{\hbar} \oint A \cdot ds = \frac{eB}{\hbar} \frac{2\pi}{n+1} \rho_T^2 \; . \tag{A.5}$$

Therefore, on using (A.4), the phase change in the equivalent "field-free" wave function $\psi_0(x)$ is

$$2\pi M' + \frac{e}{\hbar} \oint A \cdot ds = \frac{eB\rho_T^2}{\hbar} 2\pi \; . \tag{A.5a}$$

This quantity is independent of n; it agrees with the form of $\psi_0(x)$ in (3.18b), where the (total) angular momentum is

$$\Lambda = eB\rho_T^2 \; . \tag{A.5b}$$

The particular case of the uniform field B (i.e. the case $n = 1$) in (3.8b) agrees on writing ρ_0 for ρ_T.

Thus the equivalent field-free wave function $\psi_0(x)$ has the same form for any integer n. The quantisation relation (A.4) will give (in general) somewhat different spectra of values of ρ_T for different n. The condition can also be written,

$$M' = n\Phi(\rho_T)/\phi_e \qquad (M' \text{ integer}) \; ,$$

where $\Phi(\rho_T)$ is the flux through $\rho = \rho_T$.

Consider the extreme cases. If n is very large, the field B_z in (A.1) rises steeply through the value B as ρ passes ρ_T. The AB phase goes to zero (as $n \to \infty$), and only $\psi_A(x)$ is left.

For $n = 0$ we have the betatron case. M' is zero, so the whole phenomenon is in the AB phase term.

Cyclotron Oscillations. It follows from (A.3b) that

$$(2m)^{-1} \left(\rho^{-1}L_z + eA_\theta \right)^2 \psi$$

$$= \frac{(eB)^2}{2m} \left[\rho_T + n(\rho - \rho_T)^2/2\rho_T + (\text{small terms}) \right]^2 \psi \tag{A.6}$$

where the "small terms" are of order $(\rho - \rho_T)^3$ and higher. Analogous to (3.9) the radial motion is given by the equation

$$-\frac{\hbar^2}{2m}\frac{1}{\rho}\frac{\partial}{\partial\rho}\left(\rho\frac{\partial\psi}{\partial\rho}\right)+\frac{(eB)^2}{2m}$$
$$\times\left[n(\rho-\rho_T)^2+(\text{small terms})\right]\psi=E'\psi\ ,\qquad\qquad\text{(A.6a)}$$

E' being a (positive) constant. The oscillations in ρ about ρ_T have the basic frequency

$$\omega_{(n)}=n^{1/2}\omega_B\ .\qquad\qquad\qquad\qquad\qquad\qquad\text{(A.6b)}$$

The energy eigenvalues of these oscillations are

$$E'_{(n)}=(n'+1/2)n^{1/2}\hbar\omega_B\qquad(n'=0,1,2,\ldots)\ ,\qquad\text{(A.6c)}$$

and n is given in (A.1)

In Sect. 4.3 [and in (B.5) of Appendix B] it is shown that the classical theory gives the same basic frequency $\omega_{(n)}$ as in (A.6b).

So long as the wave packet remains on the circle $\rho=\rho_T$ it moves in induction of strength B, and it is clear that the frequency of rotation (ω_B) should be the same as in the uniform field case. Oscillations in ρ bring the electron into different fields, so it is not surprising that $\omega_{(n)}$ of (A.6b) differs from ω_B for $n\neq1$. In the betatron case ($n=0$) there are no such oscillations.

Appendix AII.
The Eigenstates of the Operator H_0

This appendix uses the quantum extremum method to estimate the operator H_0 that remains after the anholonomic term in the Hamiltonian is removed explicitly, as was explained in Sects. 8.10 and 8.11. This is carried out for the same axially symmetric fields as appeared in Appendix AI. The result $\mathrm{Min}(H_0')$ is expressed in terms of the electron's kinetic energy, and that is used in Sect. 8.11 as a check on the method of extracting the phase γ.

In the method used in Sect. 8.10 the operator H_0 of (8.58) is diagonalized. We have

$$H_0 = -\frac{\hbar^2}{2m}\left[\frac{\partial^2}{\partial z^2} + \rho^{-1}\frac{\partial}{\partial \rho}\left(\rho\frac{\partial}{\partial \rho}\right)\right] + \frac{1}{2m}\left[\left(\frac{M\hbar}{\rho}\right)^2 + (eA_\theta)^2\right] \quad (\text{A.7})$$

with A_θ as given in (A.2). We consider eigenstates that are independent of z, and the terms involving $\partial/\partial\rho$ are left to the study of oscillations in ρ about the value of ρ in the lowest eigenstate.

The remaining part of H_0 is

$$H_0' = (2m)^{-1}F(x) \quad,$$

where

$$F(x) = \frac{A}{x} + Cx^n \quad, \qquad x \equiv \rho^2$$

and

$$A = (M\hbar)^2 \quad, \qquad C = \left(\frac{eB}{n+1}\frac{1}{\rho_{\mathrm{T}}^{n-1}}\right)^2 \quad.$$

The minimum of $F(x)$ is at $x = x_0$, where

$$x_0^{n+1} = \frac{A}{nC} \quad (\text{A.8})$$

and

$$F(x_0) = C(n+1)x_0^n \quad . \quad (\text{A.8a})$$

We can impose the condition that $x_0 = \rho_{\mathrm{T}}^2$, which is equivalent to the equation

$$Cx_0^n = \left(\frac{eB\rho_{\mathrm{T}}}{n+1}\right)^2 \quad . \quad (\text{A.8b})$$

Thus (A.8) yields

$$M\hbar = \pm \frac{n^{1/2}}{n+1} m|\omega_B|\rho_T^2 \tag{A.9}$$

(with $\omega_B = eB/m$). Both signs of M are possible because the Hamiltonian H_0 cannot determine the sense of the motion on the circle $\rho = \rho_T$.

Thus the extremum value of H_0' is

$$\begin{aligned} \operatorname{Min}(H_0') &= \frac{1}{2}m\frac{(\omega_B\rho_T)^2}{n+1} \\ &= \frac{1}{2}m\frac{v_T^2}{n+1} \ , \end{aligned} \tag{A.10}$$

where $v_T = |\omega_B|\rho_T$ is the electron's speed.

By (8.69b) the 'anholonomic' energy is

$$E_{AN} = \frac{M\hbar|\omega_B|}{n+1} \ . \tag{A.11}$$

Using the plus sign in (A.9) gives the total energy

$$\operatorname{Min}(H_0') + E_{AN} = \frac{1}{2}mv_T^2 \left(\frac{n^{1/2}+1}{n+1}\right)^2 \tag{A.12}$$

as in (8.69c).

The minus sign in (A.9) corresponds to motion in the opposite sense, and it yields the energy

$$\operatorname{Min}(H_0') + E_{AN} = \frac{1}{2}mv_T^2 \left(\frac{n^{1/2}-1}{n+1}\right)^2 \ . \tag{A.12a}$$

This result is used in (8.69d).

Appendix B. Classical Action and General Motion in an Axially Symmetric Induction

It is shown how the action method can be applied to solve the general two-dimensional motion in a plane orthogonal to the symmetry axis in the case of the axially symmetric induction that was used in Appendix AI. The two action variables K and J_ρ are derived, and an approximation to the general solution is shown.

The classical equations of motion for an electron moving under the axial induction field $B_z(\rho)$ alone, are given in (4.3a, b). Cylindrical coordinates (z, ρ, θ) are used.

The first integral of (4.4), for the field B_z of (3.20) [or of (A.1), (A.2)] is

$$m\rho^2\dot\theta - \frac{eB}{n+1}\frac{\rho^{n+1}}{\rho_T^{n-1}} = K \quad . \tag{B.1}$$

K is a constant for a particular path. By (4.7) K is also an action variable J_θ. Define the constant α by

$$\alpha = \frac{eB}{n+1}\frac{1}{\rho_T^{n-1}} \quad . \tag{B.2a}$$

For motion on the circle $\rho = \rho'$, the field is $B(\rho')$ and the cyclotron frequency is

$$\omega(\rho') = eB(\rho')/m \quad . \tag{B.2b}$$

For motion on this circle, (B.1) gives

$$K = n\alpha(\rho')^{n+1} \quad . \tag{B.3}$$

Other paths besides this circle have the same value of K. They are distinguished by the value of the other action variable J_ρ.

Applying (B.1) twice to (4.3a) yields

$$m^2\rho^3\ddot\rho = (K + \alpha\rho^{n+1})(K - n\alpha\rho^{n+1}) \quad . \tag{B.4}$$

Using the value of K from (B.3) in (B.4) and ignoring higher terms in $(\rho - \rho')$ gives

$$\ddot\rho = -n\left(\frac{eB(\rho')}{m}\right)^2(\rho - \rho') \quad . \tag{B.5}$$

This gives the small amplitude radial oscillations about $\rho = \rho'$, with circular frequency

$n^{1/2}\omega(\rho')$.

This is in agreement with the quantum theory result in (A.6b) and (A.6c).

Radial Motion. On writing (B.4) in the form

$$\ddot{\rho} = F(\rho, K) \ ,$$

the other integral of motion is

$$\frac{\dot{\rho}^2}{2} = C + \int^{\rho} F(\rho'', K) d\rho'' \tag{B.4a}$$

where C is an arbitrary constant. Thus

$$\dot{\rho}^2 = 2C - m^{-2}(K/\rho + \alpha\rho^n)^2 \ . \tag{B.4b}$$

Using (B.1) we have

$$\rho^2\dot{\theta}^2 = m^{-2}(K/\rho + \alpha\rho^n)^2 \ . \tag{B.4c}$$

Hence with (4.4b), $C = T/m$. Equation (4.7a) becomes

$$J_\rho = \frac{m}{2\pi} \oint \left[2C - m^{-2}(K/\rho + \alpha\rho^n)^2\right]^{1/2} d\rho \ . \tag{B.6}$$

The minimum value of

$$\frac{K}{\rho} + \alpha\rho^n$$

occurs at ρ' [given by (B.3)], and it is

$$(n+1)\alpha\rho'^n \ .$$

Expanding about the minimum gives

$$(K/\rho + \alpha\rho^n)^2 = \left[(n+1)\alpha\rho'^{n-1}\right]^2 \left[\rho'^2 + n(\rho - \rho')^2\right] \ , \tag{B.6a}$$

where higher order terms are omitted. We shall use this approximation to evaluate (B.6).

Simple integration yields,

$$J_\rho \simeq \frac{m}{2} n^{-1/2} \left[\frac{v^2}{\omega(\rho')} - \rho'^2\omega(\rho')\right] \ . \tag{B.7}$$

We have written $2C = v^2$. Equation (B.7) will be accurate when $2C$ is close to the minimum of the curve in Fig. 8.13.2, but it will be less accurate for large values of $2C$.

The Hamiltonian corresponding to the Lagrangian in (4.5) is

$$H = \dot{\rho}\frac{\partial L}{\partial\dot{\rho}} + \dot{\theta}\frac{\partial L}{\partial\dot{\theta}} - L = \frac{mv^2}{2} \ . \tag{B.8}$$

Thus (B.7) expresses H in terms of J_ρ and ρ', and thus by (B.3), H is written in the form $H = H(K, J_\rho) = H(J_\theta, J_\rho)$.

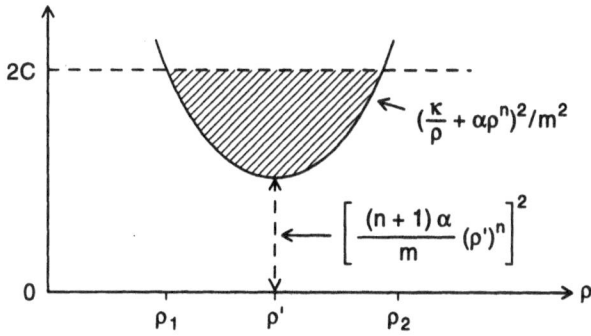

Fig. B.1. The hatched area gives the square of the integrand in (B.6)

We now evaluate (4.9). Equation (B.7) can be written

$$v^2 = \omega(\rho')^2 \rho'^2 + \frac{2n^{1/2}}{m} J_\rho \omega(\rho') \quad . \tag{B.7a}$$

It is clear from (4.7a) that $J_\rho = 0$ implies that the orbit is a circle, $\rho = $ const. In that case (B.7a) is obvious.

In general

$$\frac{m}{2} \frac{\partial v^2}{\partial K} = \omega(\rho') + n^{1/2} \left(1 - \frac{1}{n} \right) \frac{J_\rho}{m\rho'^2} \quad . \tag{B.9}$$

By (4.9) this relation is the same as

$$\dot{\phi}_\theta = \frac{\partial H}{\partial J_\theta} \quad .$$

For $J_\rho = 0$

$$\dot{\phi}_\theta = \omega(\rho') \quad ,$$

in agreement with (3.5a).

In general, by (B.7),

$$\dot{\phi}_\rho = \frac{m}{2} \frac{\partial v^2}{\partial J_\rho} = n^{1/2} \omega(\rho') \quad . \tag{B.9a}$$

This agrees with the special case in (B.5).

In certain orbits, such as that shown in Fig. 4.1, we expect to have

$$\dot{\phi}_\rho = \dot{\phi}_\theta \quad . \tag{B.10}$$

The right-hand sides of (B.9) and (B.9a) should be equal in those cases. For $n > 1$, this gives

$$v^2 = \left(1 + \frac{2n}{1 + n^{1/2}} \right) \rho'^2 (\omega(\rho'))^2 \quad , \tag{B.10a}$$

i. e.

$$\frac{J_\rho}{m} = \frac{n^{1/2}}{1 + n^{1/2}} \rho'^2 \omega(\rho') \quad . \tag{B.10b}$$

A trivial exception is the case $n = 1$, (i. e. the uniform field $\boldsymbol{B}$), for which (B.10) always holds.

The quadratic approximation used to evaluate (B.6) gives the separation of the extreme values ρ_1, ρ_2 (of ρ) in an orbit to be

$$\rho_2 - \rho_1 = 2n^{-1/4} \left(\frac{2J_\rho}{m\omega(\rho')} \right)^{1/2} . \tag{B.11}$$

Notice that in all these relations the parameter ρ' is merely a function of K, which is defined in (B.3). Thus ρ' and J_ρ specify the orbit: ρ' is the radius of an orbit only for $J_\rho = 0$. Also $\omega(\rho')$ would be the cyclotron frequency for the orbit $\rho = \rho'$.

$J_\rho \neq 0$ can describe both (i) the orbits of the type shown in Fig. 4.1 and (ii) the radial oscillations given by (4.7b). Thus e. g. using the example in Sect. 4.2, and putting $\omega = n^{1/2}\omega(\rho')$ in (4.7b), we see that (B.11) gives $\rho_2 - \rho_1 = 2D$.

This completes our account of how to write H in terms of the action variables $K (= J_\theta)$, and J_ρ in the case of the field $\boldsymbol{B} = (0, 0, B_z)$ where B_z is of the form in (3.20). For an account of the use of action variables in problems concerning electrons moving in more general magnetic fields, see for example [B.1], [B.2].

Appendix C. Solutions for the Case of the Toroidal Solenoid

Orthogonal coordinates are set up for the toroidal system and the metric properties are examined. A vector potential describing the solenoid induction is given and special relations are pointed out which make it easy to see solutions of Schrödinger's equation for an electron circulating around and through the torus. The similarities with the treatment of the motion around a linear solenoid in Chap. 2 are pointed out, including the centrifugal potential term and a circulation operator.

A brief survey of the formalism that leads to key relations similar to (2.12) and (2.13) is given. The uniformly wound solenoid, and a typical orbit around it, are illustrated in Fig. C.1.

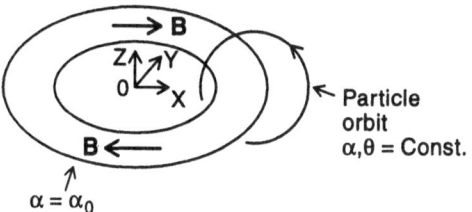

Fig. C.1. Toroidal solenoid and the path of a particle

Cartesian coordinates (x, y, z) and cylindrical polars (z, ρ, θ) are related to toroidal coordinates (α, β, θ) as follows:

$$\rho = \frac{k}{\chi} \sinh \alpha \ , \qquad z = \frac{k}{\chi} \sin \beta \ , \qquad \frac{y}{x} = \tan \theta \ , \tag{C.1}$$

with

$$\chi = \cosh \alpha - \cos \beta \ .$$

k is a scale length and the whole space is

$$0 \leq \alpha < \infty \ , \qquad 0 \leq \beta \leq 2\pi \ , \qquad 0 \leq \theta \leq 2\pi \ .$$

The distance from the origin is

$$|x| = k \left(\frac{\chi'}{\chi} \right)^{1/2} \ , \tag{C.2}$$

where

$$\chi' = \cosh\alpha + \cos\beta \ .$$

It follows that $\chi' = 0$ (i.e. $\alpha = 0$, $\beta = \pi$) is the origin $x = 0$. By (C.1), $\alpha = 0$, $|\beta| > 0$ is the axis Oz. In the neighbourhood of $\chi = 0$ (i.e. $\alpha = 0$, $\beta = 0$) we have

$$|x| = \frac{2k}{(\alpha^2 + \beta^2)^{1/2}} \ , \qquad \frac{z}{\rho} = \frac{\beta}{\alpha} \ , \qquad \frac{y}{x} = \tan\theta \ . \tag{C.3}$$

Thus $\alpha \to 0$, $\beta \to 0$ gives $|x| \to \infty$. This is a singular point in the coordinate system. We shall stay on the sheet with $\chi^{1/2} > 0$.

There are two orthogonal families of circles lying in any plane $\theta = $ const.: (cf. Fig. C.2)

$$\alpha = \text{const.:} \quad (\rho - k\coth\alpha)^2 + z^2 = \left(\frac{k}{\sinh\alpha}\right)^2 \ ,$$

$$\beta = \text{const.:} \quad \rho^2 + (z - k\cot\beta)^2 = \left(\frac{k}{\sin\beta}\right)^2 \ . \tag{C.4}$$

The torus is $\alpha = \alpha_0$, $0 \le \theta \le 2\pi$. It has internal radius $r_T = k/\sinh\alpha_0$, and the radius of centres is $R = k\coth\alpha_0$. A particle orbit $\alpha = $ const. (with $\alpha < \alpha_0$), $\theta = $ const. is shown in Fig. C.2

The limit $\alpha \to \infty$ describes the circle $\rho = k$, $z = 0$; we shall call this the limit circle. For large α we have

$$\rho - k \simeq \frac{k\cos\beta}{\chi} \ , \tag{C.5}$$

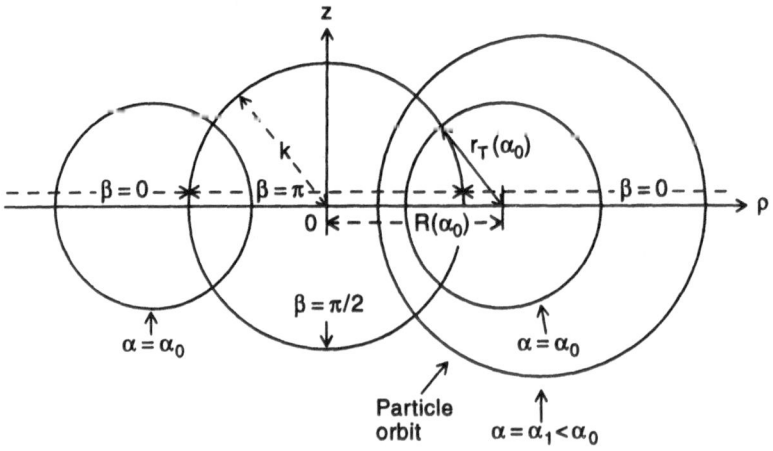

Fig. C.2. The orthogonal circles $\alpha = $ const. and $\beta = $ const. in a section $\theta = $ const. through Oz

and using (C.1) it follows that for large α the representative point is at a distance

$$D \simeq \frac{k}{\chi} \tag{C.5a}$$

from the limit circle. For fixed large α, on varying β from 0 to 2π, the point moves around the limit circle (in any plane $\theta = $ const.) in a manner that is important for tracing the centrifugal force term in what follows.

Differential Properties. The element of length is

$$ds = (h_1\, d\alpha,\ h_2\, d\beta,\ h_3\, d\theta)\ ,$$

with

$$h_1 = h_2 = \frac{k}{\chi}\ , \qquad h_3 = \frac{k}{\chi}\sinh\alpha\ . \tag{C.6}$$

Assume that inside the solenoid $\alpha = \alpha_0$ the induction $\boldsymbol{B}$ is everywhere along the direction θ, while $\boldsymbol{B} = 0$ outside (i.e. for $\alpha < \alpha_0$). Also the flux in the solenoid is Φ in each section $\theta = $ const. A suitable vector potential, in $\alpha < \alpha_0$, is

$$\boldsymbol{A} = (0, A_B, 0)\ , \qquad A_B = \frac{\Phi\chi}{2\pi k}\ . \tag{C.7}$$

This follows from integrating around any orbit encircling the torus in $\alpha < \alpha_0$ which gives:

$$\oint \boldsymbol{A}\cdot d\boldsymbol{s} = \oint A_B h_2\, d\beta = \Phi\ . \tag{C.7a}$$

Equations (C.2) and (C.3) show that A_B varies as $|\boldsymbol{x}|^{-2}$ for large $|\boldsymbol{x}|$. Also since $h_2 A_B$ is a constant, curl $\boldsymbol{A}$ vanishes in $\alpha < \alpha_0$.

Standard formulae give (for $\alpha < \alpha_0$):

$$\text{div}\,\boldsymbol{A} = \frac{1}{h_1 h_2 h_3}\frac{\partial}{\partial\beta}(h_1 h_3, A_B) = -\frac{\Phi}{2\pi}\frac{\chi\sin\beta}{k^2}\ , \tag{C.8a}$$

$$\boldsymbol{A}\cdot\text{grad}\,\psi = \frac{\Phi}{2\pi}\left(\frac{\chi}{k}\right)^2\frac{\partial\psi}{\partial\beta}\ . \tag{C.8b}$$

Several relations are useful in setting up solutions of Schrödinger's equation. There is the equation

$$k^2\boldsymbol{\nabla}^2\chi^{1/2} = \tfrac{1}{4}\chi^{5/2}\ . \tag{C.9}$$

Also using (C.7), (C.8a) and (C.8b) gives the form

$$k^2(\text{i}\boldsymbol{\nabla} + \boldsymbol{A})^2\psi = \chi^2\psi\left[\left(n - \frac{\Phi}{2\pi}\right)^2 - \frac{1}{4}\right]\ , \tag{C.9a}$$

where ψ is the wave function

$$\psi = \chi^{1/2}\exp(\text{i}n\beta)\ , \qquad (n = \text{integer})\ . \tag{C.9b}$$

By (C.2)

$$\chi^{1/2} \simeq 2^{1/2} \frac{k}{|x|} \quad , \qquad \text{as} \quad \chi \to 0 \quad ,$$

so $\psi = \chi^{1/2}$ is not a normalisable wave function (as $|x| \to \infty$). However we can use the form

$$\psi = \chi^{1/2} g(\alpha, \beta, \theta) \quad , \tag{C.10a}$$

and require that

$$g \to 0 \quad \text{as} \quad |x| \to \infty \quad .$$

A further advantage in using this form is that Schrödinger's equation for g becomes separable. In fact, (C.10a) yields

$$k^2 \nabla^2 \psi = \chi^{5/2} \left[\frac{1}{\sinh \alpha} \frac{\partial}{\partial \alpha} \left(\sinh \alpha \frac{\partial g}{\partial \alpha} \right) \right.$$
$$\left. + \frac{\partial^2 g}{\partial \beta^2} + \frac{g}{4} + \frac{1}{\sinh^2 \alpha} \frac{\partial^2 g}{\partial \theta^2} \right] \quad . \tag{C.10b}$$

Application. For an electron, by using (C.8a, b), we get

$$i\hbar \frac{\partial g}{\partial t} = \frac{\hbar^2 \chi^2}{2mk^2} \left\{ - \left[\frac{1}{\sinh \alpha} \frac{\partial}{\partial \alpha} \left(\sinh \alpha \frac{\partial g}{\partial \alpha} \right) + \frac{1}{\sinh^2 \alpha} \frac{\partial^2 g}{\partial \theta^2} \right] \right.$$
$$\left. + \left[\left(\frac{\partial}{i \partial \beta} + P(t) \right)^2 - \frac{1}{4} \right] g \right\} \quad . \tag{C.11}$$

A potential $V(\alpha)$ can be added to the Hamiltonian in order to confine the orbit of the electron to a range of values of α. The last term in the curly brackets in (C.11) is the "centrifuge potential" term [this follows from the presence of χ^2 and the result in (C.5a)]. $P(t)$ is defined in (2.10a), and (C.11) should be compared with (2.12) in Sect. 2.4.

We choose a single-valued wave function of the form

$$\psi = f(\alpha, t) \chi^{1/2} \exp(iM\beta) \quad , \qquad (M = \text{integer}) \quad . \tag{C.11a}$$

The equation for $f(\alpha, t)$ is

$$i\hbar \frac{\partial f}{\partial t} = -\frac{\hbar^2 \chi^2}{2mk^2} \frac{1}{\sinh \alpha} \frac{\partial}{\partial \alpha} \left(\sinh \alpha \frac{\partial f}{\partial \alpha} \right)$$
$$+ \frac{\hbar^2 \chi^2}{2mk^2} \left\{ [M + P(t)]^2 - \frac{1}{4} \right\} f \quad , \tag{C.11b}$$

and the boundary condition is

$$f(\alpha, t) \to 0 \quad , \qquad \alpha \to 0 \quad . \tag{C.11c}$$

Centrifugal Potential. As in Sect. 2.4, the centrifugal potential term

$$\frac{\hbar^2 \chi^2}{2mk^2} \left[M + P(t) - \frac{1}{2} \right] \left[M + P(t) + \frac{1}{2} \right] \tag{C.12}$$

gives the basis for the physical interpretation. Equation (C.12) is to be compared with (2.12a) in Sect. 2.4.

Circulation Operator. An operator S_B represents the circulation around the torus $\alpha = \alpha_0$. For the type of wave function in (C.10a), we have

$$S_B \equiv \frac{\hbar}{i} \chi^{1/2} \frac{\partial}{\partial \beta} \chi^{-1/2} = \frac{\hbar}{i} \left(\frac{\partial}{\partial \beta} - \frac{1}{2} \chi^{-1} \sin \beta \right) \quad .$$

Analogous to (2.13), the operator

$$\sum_B = \chi^{1/2} \left(\frac{\hbar}{i} \frac{\partial}{\partial \beta} + \hbar P(t) \right) \chi^{-1/2}$$

will describe the total circulation when the potential A_B of (C.7) is present. There are special solutions of (C.11). These are of the form

$$g = f(\alpha, t) \cos(\beta/2) \quad \text{and} \quad f(\alpha, t) \sin(\beta/2) \quad .$$

For $P(t) = 0$, these terms have zero centrifugal potential, by (C.12). These are not single-valued solutions. In the first solution, on increasing β by 2π (on α, $\theta = $ const.), the solution reverses sign. In the second case the derivative $\partial\psi/\partial\beta$ reverses sign on increasing β by 2π.

For information on using toroidal coordinates see, for example, [C.1].

$$\left[\int_0^\infty dr \, r^2 \varphi_i(r)\right] = \int_0^\infty dr \, r^2 \varphi_i(r) \frac{1}{2\pi^2}$$

$$\frac{\partial}{\partial t} \cdot \cdots \cdot \int \cdots \left(\frac{\partial}{\partial r} - \cdots\right) \cdots$$

Appendix D. Angle and Action
for the Simple Harmonic Oscillator

This appendix shows the simplest case of an angle and action pair of canonical conjugate variables.

For the simple harmonic oscillator having mass m and circular frequency ω the Hamiltonian is

$$H = \frac{p^2}{2m} + m\omega^2 \frac{q^2}{2} \quad . \tag{D.1}$$

Thus

$$p = \left[2m(W - m\omega^2 q^2/2)\right]^{1/2} \quad ,$$

where the energy W is an arbitrary positive parameter. The action is

$$J = \frac{1}{2\pi} \oint p \, dq \quad , \tag{D.2}$$

where the integral is taken over one cycle. Integration gives

$$J = \frac{W}{\omega} \quad . \tag{D.2a}$$

The Hamilton–Jacobi function $S(q, J)$ obeys

$$\frac{\partial S(q, J)}{\partial q} = p \quad ,$$

so

$$S(q, J) = (2m\omega)^{1/2} \int_0^q \left(J - m\omega^2 q'^2/2\right)^{1/2} dq' \quad . \tag{D.3}$$

The angle variable conjugate to J is now given by

$$w(q, J) = \left(\frac{\partial S(q, J)}{\partial J}\right)_q$$

$$= \sin^{-1}(q/q_0) \quad ; \qquad q_0 = (2J/m\omega)^{1/2} \quad . \tag{D.4}$$

The angle variable w increases steadily as q oscillates backwards and forwards between $+p_0$ and $-q_0$. The root

$$\left(J - mq'^2/2\right)^{1/2}$$

in the integrand in (D.3) reverses sign each time that q reaches q_0 or $-q_0$, so $S(q, J)$ increases (or decreases) steadily during the motion.

The general equations and (D.2a) yield

$$\dot{w} = \frac{\partial W}{\partial J} = \omega \quad .$$

Now (D.4) gives

$$q = q_0 \sin w = q_0 \sin(\omega t + \varepsilon) \quad , \tag{D.4a}$$

where ε is a constant. Inserting (D.4a) in (D.3) yields

$$S(q, J) = 2J \int_0^w \cos^2 w' \, dw'$$
$$= J(w + \sin 2w) \quad . \tag{D.5}$$

Thus S equals Jw plus a function periodic in $2w$. That property is important for the adiabatic theorem in Sect. 8.3.

S can be written as a function of q and J, or as a function of w and J. In the latter case one should remember that $w = w(q, J)$. In general

$$\left(\frac{\partial S(q, J)}{\partial J}\right)_q = \left(\frac{\partial S(w, J)}{\partial J}\right)_w + \left(\frac{\partial S(w, J)}{\partial w}\right)_J \left(\frac{\partial w}{\partial J}\right)_q \quad . \tag{D.6}$$

For the simple harmonic oscillator, using (D.4a) and remembering that q_0 depends on J, as in (D.4), we find that

$$\left(\frac{\partial w}{\partial J}\right)_q = -\frac{1}{2J} \tan w \quad . \tag{D.6a}$$

On using (D.5) it is now easy to verify that the left-hand side of (D.6) is w itself, as is required by (8.9).

Appendix E. Angle and Action for a General Oscillator

This appendix shows an example of a generalisation of the variables appearing in Appendix D.

An extension of the simple harmonic oscillator (Appendix D) is given by the Hamiltonian

$$H = \frac{p^2}{2m} + \lambda V(q) \ , \tag{E.1}$$

where $V(q)$ has the properties

$$V(-q) = V(q) \ , \qquad V(0) = 0 \ , \qquad \frac{d^2 V(q)}{dq^2} > 0 \ ,$$

and also $\lambda > 0$.

The Hamilton–Jacobi function is $S(q, J)$ which is given by

$$S(q, J) = (2m)^{1/2} \int_0^q [W - \lambda V(q')]^{1/2} \, dq' \ . \tag{E.2}$$

For convenience we shall write

$$Q(q') = [W - \lambda V(q')]^{1/2} \ . \tag{E.2a}$$

By definition the action variable is

$$J = \frac{1}{2\pi} \oint p \, dq = \frac{2(2m)^{1/2}}{\pi} \int_0^{q_0} Q(q') \, dq' \ , \tag{E.3}$$

where q_0 is defined to be positive, and satisfies

$$\lambda V(q_0) = W \ . \tag{E.3a}$$

This gives $q_0 = q_0(\lambda, W)$. The form in (E.3) is due to the cycle being $0 \to q_0 \to -q_0 \to 0$. That gives four times the integral over the range $(0, q_0)$.. Also $J = J(W)$.

Equation (E.3) and (E.3a) express J as a function of W (keeping λ etc. constant). Also

$$\frac{dJ}{dW} = \frac{1}{\pi}(2m)^{1/2} \int_0^{q_0} \frac{dq'}{Q(q')} > 0 \ .$$

It follows that

$$w = \frac{\partial S(q, J)}{\partial J} = \frac{\dfrac{\pi}{2} \displaystyle\int_0^q \dfrac{dq'}{Q(q')}}{\displaystyle\int_0^{q_0} \dfrac{dq'}{Q(q')}} \ . \tag{E.4}$$

Equation (E.4) defines

$$w = w(q, J) \ ,$$

with $w = 0$ for $q = 0$, and $w = \pi/2$ for $q = q_0$.

Equation (E.2) can now be written as

$$S - wJ = \frac{\pi J}{2} \left(\frac{\displaystyle\int_0^q Q(q')\,dq'}{\displaystyle\int_0^{q_0} Q(q')\,dq'} - \frac{\displaystyle\int_0^q (Q(q'))^{-1}\,dq'}{\displaystyle\int_0^{q_0} (Q(q'))^{-1}\,dq'} \right) . \tag{E.5}$$

The corresponding value of w and q are:

w:	0,	$\pi/2$	π,	$3\pi/2$,	2π,	$5\pi/2$,	3π
q:	0,	q_0	0,	$-q_0$,	0,	q_0,	0
Sign of Q:	+,	0,	$-$,	0,	+,	0,	$-$.

$Q(q')$ is given by (E.2a). As $Q(q')$ passes through 0 it changes from one branch of the square root to the other. The signs of Q are shown above. Consequently each of the integrals

$$\int_0^q Q(q')\,dq' \qquad \text{and} \qquad \int_0^q (Q(q'))^{-1}\,dq'$$

increases steadily when w is increased.

It follows that the term in the curly brackets in (E.5) vanishes at $w = 0$, $\pi/2$, π, $3\pi/2$, 2π and so on. This term is the difference of two continuous functions of w, so it itself is continuous, and will change sign at its zeros. Hence

$$S = wJ + S^+(w, J) \ , \tag{E.6}$$

where $S^+(w, J)$ is periodic in w with period π. Also S^+ is an odd function of w.

$S^+(w, J)$ can be expressed in a Fourier series of the form

$$S^+(w, J) = \sum_{n=1}^{\infty} b_n \sin(2nw) \ . \tag{E.7}$$

The coefficients b_n depend on J.

General Form of $S^+(w, J)$. For the general case of cyclic or periodic motion at a constant value of J, it follows from (8.10) that w increases linearly with t. By (8.9)

$$w = \left(\frac{\partial S(q, J)}{\partial J} \right)_q . \tag{E.8}$$

It follows that S itself must contain a term that is linear in t.

On using (D.6) it is seen that the form of $S(w, J)$ in (E.6) will give rise to the correct left-hand side of (E.8), provided that $S^+(w, J)$ obeys

$$\left(\frac{\partial S^+(w, J)}{\partial J}\right)_w + \left[J + \left(\frac{\partial S^+(w, J)}{\partial w}\right)_J\right]\left(\frac{\partial w}{\partial J}\right)_q = 0 \ . \tag{E.9}$$

If (E.9) is obeyed, then by (E.6)

$$\left(\frac{\partial S^+(q, J)}{\partial J}\right)_q + J\left(\frac{\partial w(q, J)}{\partial J}\right)_q = 0 \ ,$$

and (E.8) will be valid. Equations (D.5) and (D.6) provide a simple example of how (E.9) can work.

Appendix F. The Parallel Transport of Vectors

This appendix gives come simple examples of the parallel transport of vectors. The cases chosen are parallel transport along a curve of vectors tangential to a smooth surface and parallel transport of vectors orthogonal to a twisted curve.

The idea of the parallel transport of a vector may not be easy to envisage, and for this reason we shall discuss a couple of examples.

Parallel transport of a vector v in three-dimensions involves no difficulty, but parallel transport relative to a surface or relative to a curve may be harder to understand. These cases are: (a) parallel transport of vectors tangential to a smooth surface along a prescribed curve on the surface. This case is related to Foucault's pendulum Sect. 8.5; (b) parallel transport along a twisted curve of vectors that are constrained to be orthogonal to the curve. This is related to the rotation of a photon passing along a twisted light pipe.

Our purpose here is merely to make clear the mathematical concept in the two cases.

(a) Parallel Transport Relative to a Surface

The simple example concerns a circle C on a sphere S. C is not a great circle (Fig. F.1).

D is a fixed point such that all planes that are tangential to S at any point of C contain the point D.

Points P and P' lie on C and are fairly close together. PD is a tangent to S and for the moment we denote it by v. $P'D$ is also a tangent to S, but it is not parallel to v. This is the basic feature of the problem.

Let v' be a tangent vector "parallel" to v. By virtue of the statement in Sect. 8.4 [just before (8.21)], the angle between v and v' will vanish quicker than ds as $ds \rightarrow 0$. If C has latitude α and $d\phi$ is the difference in longitude of P and P', the angle between v and v' must vanish faster than $d\phi$ as $d\phi \rightarrow 0$.

The angle between v' and $P'D$ is

$$d\psi = -\frac{ds}{L}$$

where $L = PD$. Thus

$$d\psi = -d\phi \sin \alpha \quad . \tag{F.1}$$

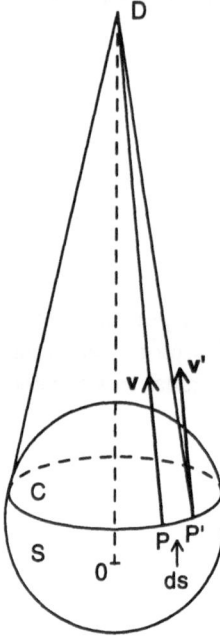

Fig. F.1. Parallel tangent vectors on a minor circle C on the sphere S

Hence for small $d\phi$, the angle $d\psi$ gives the amount by which the parallel vector v' has rotated beyond $P'D$. (The sign conventions can of course be altered by looking at the diagram in another way.)

If u is another vector at P which is tangential to S and u' is its parallel tangent vector at P', then we would expect that the angle δ between v and u will equal the angle δ' between v' and u', when $d\phi$ is small. It is shown in [F.1] pp. 178–180, that the precise condition for parallelism (i. e. the vanishing of the surface intrinsic derivative along C of the contravariant tangent vector) will ensure that (a) $\delta' = \delta$ always and (b) the lengths of v and v' are always equal.

It now follows that $d\psi$ in (F.1) gives the rotation of any tangent vector u (relative to the line PD) on parallel transport along C through the longitude $d\phi$.

Adding the effect in (F.1) over numerous small steps $d\phi$, it follows that if a tangent vector v is moved around the circle C until it again reaches its starting point, the vector will still be tangential to S, but it will lie at the angle

$$\psi = -2\pi \sin \alpha \qquad \text{(F.2)}$$

relative to its original position.

A similar definition of parallel transport applies to surfaces S other than the sphere. The definitions all relate to transport along a given curve C on a surface S. Without a curve C the problem is not determined. Detailed references are given in Sect. 8.4. There are interesting general results; for example, for parallel transport along a geodesic, there is no rotation of the tangential vector v. The case $\alpha = 0$ in (F.2) is an example of this property, since $\alpha = 0$ is a great circle.

Another general property is seen in (8.21) of Sect. 8.4, namely that the surface S must have non-zero curvature (Gauss curvature). Otherwise there will be no interesting properties. Equation (8.21) expresses the angle

$$\beta = 2\pi + \psi \tag{F.2a}$$

in terms of the Gaussian curvature of the portion of the surface S within the closed curve C. In the case where S is a sphere, β is the solid angle subtended at the centre by C.

For $\alpha \simeq \pi/2$ in Fig. F.1 we have

$$\psi \simeq \pm 2\pi \ .$$

This applies to an observation made near the Earth's S-pole or N-pole respectively. The details are left as an exercise.

The relation between β and ψ given in (F.2a) shows that they are the inward and outward measurements of the angle between two vectors, both measured in the same sense.

It is clear from Sect. 8.5 and the present description, that the simplest explanation of Foucault's pendulum is the parallel transport of tangential vectors. The details of the dynamics are redundant.

For recent textbook accounts of parallelism on a surface see [F.2], [F.3]. In particular for a simple discussion of the various differential geometry properties that are associated with parallelism on a surface see Chap. 12 of [F.3].

Example on the Sphere. Various properties that are mentioned above may be more lucid if the following notes on transport on the spherical surface are used.

On the sphere in Fig. F.1 the origin of polar coordinates is O, and any point P on the surface is given by

$$r = (\sin \theta \cos \phi, \sin \theta \sin \phi, \cos \theta) \ , \qquad (0 \leq \theta \leq \pi, \ 0 \leq \phi \leq 2\pi) \ .$$

Unit tangent vectors at P are

$$i_\theta = (\cos \theta \cos \phi, \cos \theta \sin \phi, -\sin \theta) \ ,$$
$$i_\phi = (-\sin \phi, \cos \phi, 0) \ .$$

The vectors i_θ and i_ϕ are orthogonal; they correspond to the infinitesimal changes in coordinates $\delta\theta$ and $\delta\phi$ respectively. The changes in the vectors i_θ and i_ϕ themselves due to changes $\delta\theta$, $\delta\phi$ are

$$\delta i_\theta = -r\delta\theta + \cos \theta i_\phi \delta\phi \ ,$$
$$\delta i_\phi = -\cos \theta i_\theta \delta\phi - \sin \theta r \delta\phi \ .$$

Only the components of these changes that lie in the tangent plane at the current point (θ, ϕ) are used in discussing tangent vectors. They will be labelled $\delta_T i_\theta$, etc.

If a, b are any real numbers then

$$\delta_T (a i_\theta + b i_\phi) = (a i_\phi - b i_\theta) \cos \theta \delta \phi \ .$$

Notice that

$$(a i_\theta + b i_\phi) \cdot (a i_\phi - b i_\theta) = 0 \ .$$

Thus under $\theta \to \theta + \delta\theta$, $\phi \to \phi + \delta\phi$, the change δ_T rotates the tangent plane at P, about the normal, through the angle

$$\cos\theta\delta\phi \ ;$$

and for $\cos\theta > 0$ the rotation is in the direction from i_θ to i_ϕ. (We have to emphasise that motion of tangent vectors along r, i. e. normal to the tangent plane, is ignored in forming δ_T.)

Now choose a closed curve C on the sphere and make a succession of changes $(\delta\theta, \delta\phi)$ to pass around C.

(i) If C is a curve $\theta = $ const. then the tangent plane rotates by

$$\cos\theta\delta\phi$$

under $\delta\phi$. If vectors v at (θ, ϕ) and v' at $(\theta, \phi + \delta\phi)$ are parallel then v' has rotated through

$$- \cos\theta\delta\phi$$

relative to the rotating tangent plane (i_θ, i_ϕ). This agrees with (F.1) and (F.2) above.

(ii) If C is a curve $\phi = $ const. the change δ_T does not rotate the tangent plane, and v, v' stay parallel as we pass around C. As $\phi = $ const. is a great circle, this result agrees with the general statement above about parallel transport along any geodesic curve.

(iii) On any closed curve C on the sphere S, the vector v' is rotated through the total angle

$$- \int_C \cos\theta \, d\phi = \int_C A_\phi \, ds_\phi \ .$$

Here A_ϕ is regarded as the ϕ-component of a surface vector and $ds_\phi = \sin\theta \, d\phi$ is the ϕ-component of ds. Thus

$$A_\phi = - \cot\theta \ ,$$

and the normal component of curl A is

$$(\text{curl } A)_r = \frac{1}{\sin\theta} \frac{\partial}{\partial\theta} (A_\phi \sin\theta) = +1 \ .$$

By Stoke's theorem it follows that the total rotation of v' is

$$\int_C dS \ ,$$

namely the surface area enclosed by C. This agrees with the angle β in (8.21).

(b) Parallel Transport of Orthogonal Vectors Along a Twisted Curve

A twisted curve C is described by the unit vectors $t(s)$, $n(s)$, $b(s)$. This orthogonal triad of tangent, normal and binormal, is described in Sect. 8.12, and s is the arc length, which increases in the direction of t.

The Serret–Frenet formulae are

$$\frac{dt}{ds} = kn \ , \qquad \frac{db}{ds} = -\tau n \ , \qquad \frac{dn}{ds} = \tau b - kt \ , \tag{F.3}$$

where k is the curvature and τ the torsion at point s.

Any vector v orthogonal to C at P will lie in the plane $\mathcal{P}$ indicated in Fig. F.2. As s alters a little, P moves along C and the plane $\mathcal{P}$ will, in general, twist around a little. What vector v' in the new position of the plane, is parallel to v?

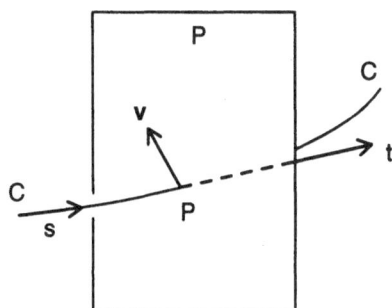

Fig. F.2. The plane $\mathcal{P}$ orthogonal to curve C at point P

If the torsion is everywhere zero:

$$\tau(s) = 0 \ , \tag{F.4}$$

then by (F.3), b will be constant. The curve C will lie in a plane orthogonal to b. Any vector v orthogonal to C and parallel to b stays trivially parallel to itself. Any vector orthogonal to b and parallel to n is parallel transported around the curve C like n. On passing around a closed curve C the triad (t, n, b) is rotated through 2π.

Consider the case that $\tau(s)$ is not zero. Any vector v lying in the plane $\mathcal{P}$ in Fig. F.2 can be expressed in the form

$$v(s) = a(s)n + c(s)b \tag{F.5}$$

where $a(s)$ and $c(s)$ are real numbers. By (F.3)

$$\frac{dv}{ds} = \frac{da(s)}{ds}n + \frac{dc(s)}{ds}b + \tau a(s)b - \tau c(s)n - ka(s)t \quad . \tag{F.5a}$$

We use an argument similar to that given in section (a) of this appendix. In order to keep $v(s)$ in the plane $(n(s), b(s))$, the last term on the right of (F.5a) is dropped. The vector is now denoted by $v'(s)$. The condition that v' stays parallel to its initial value v requires

$$\frac{dv'}{ds} = 0 \quad . \tag{F.6}$$

This yields

$$\frac{da(s)}{ds} = \tau c(s) \quad , \qquad \frac{dc(s)}{ds} = -\tau a(s) \quad .$$

Hence

$$\frac{d}{ds}[a(s) + ic(s)] = -i\tau(s)[a(s) + ic(s)] \tag{F.6a}$$

so

$$a(s) + ic(s) = \exp\left[-i\int_0^s \tau(s')\,ds'\right][a(0) + ic(0)] \quad . \tag{F.7}$$

Equations (F.5) and (F.7) describe the parallel transport of vectors v' orthogonal to C. The vector $v'(s')$ is rotated in the (moving) plane orthogonal to C by the angle

$$-\int_0^s \tau(s')\,ds' \tag{F.8}$$

on passing from $s' = 0$ to $s' = s$. In order to use this relation it is desirable that $(n(s'), b(s'))$ have the same directions at $s' = 0$ and $s' = s$. The triad (t, n, b) has to have the same orientation at the two ends of the portion of twisted curve that is used.

The Circular Helix. For a circular helix with radius a the position is

$$r = u(\cos\theta, \sin\theta, \theta\tan\alpha) \quad ,$$

where α is constant. Then the curvature and torsion are

$$k = \cos^2\alpha/a \quad , \qquad \tau = \cos\alpha\sin\alpha/a \quad .$$

Also,

$$n = (-\cos\theta, -\sin\theta, 0) \quad ,$$
$$b = \sin\alpha(\sin\theta, -\sin\cos\theta, \cot\alpha) \quad ,$$
$$t = (-\sin\theta\cos\alpha, \cos\theta\cos\alpha, \sin\alpha) \quad .$$

The length of one circuit $(0 \leq \theta \leq 2\pi)$ is

$$L = \frac{2\pi a}{\cos\alpha}$$

and (F.8) becomes

$$-L\tau = -2\pi \sin\alpha \ .$$

This is the total angle of rotation of the vector $v'(s')$ in the plane $\mathcal{P}$ in one turn of the circular helix.

If its base is kept fixed at O, the (unit) vector b describes a circle of radius $\sin\alpha$ that lies in the plane $z = \cos\alpha$. In this case the orthogonal tangent vector t in one circuit describes a cone having semi-angle $(\pi/2 - \alpha)$. The area of a unit sphere enclosed by this orbit is

$$2\pi(1 - \sin\alpha) \ .$$

(c) General Theory

In differential geometry a Lie group, or a Lie algebra, can provide a geometrical connection and thereby define parallel displacement of geometrical entities. There is a close relationship to the properties of non-Abelian gauge fields. For a concise survey, see Chap. 12 of Ref. [F.4].

Appendix G. The Basic Solution and Aharonov–Bohm Scattering

This appendix discusses and compares solutions (i) for the incident electron plane wave passing an infinite straight flux line of zero radius, giving the basic solution $\psi_B(\rho, \theta)$, which shows the simplest AB effect; and (ii) for the similar case but with the boundary condition that the wave function ψ vanishes at the flux line – this gives the AB scattering solution $\psi_{AB}^P(\rho, \theta)$. (Cylindrical polar coordinates are used.)

The mathematical behaviour of the two solutions is discussed, and two-dimensional and three-dimensional scattering are compared.

Consider the limiting case in which the induction B in Fig. 2.1, or Fig. 2.3 is confined to a line (i. e. O_z). This means that the radius of the cylinder ρ_C tends to zero. At the same time B is increased so that the flux Φ is kept constant. Suppose also that there is no confining potential, and the electron moves freely in the vector potential

$$\boldsymbol{A} = (0, 0, A_\theta) \ , \qquad A_\theta = \frac{\Phi}{2\pi\rho} \ , \qquad (\rho > 0) \ . \tag{G.1}$$

If the electron moves with a propagation vector $\boldsymbol{k}$ that is along O_x and the wave function is independent of z, then in cylindrical coordinates Schrödinger's equation (2.10) becomes

$$\frac{\partial^2\psi}{\partial\rho^2} + \frac{1}{\rho}\frac{\partial\psi}{\partial\rho} + \left[k^2 + \frac{1}{\rho^2}\left(\frac{\partial}{\partial\theta} + i\alpha\right)^2\right]\psi = 0 \ , \qquad (\rho > 0) \ , \tag{G.2}$$

where

$$\alpha = \frac{\Phi}{\phi_e} \ . \tag{G.2a}$$

(It should be remembered that the electron's charge is $-e$.)

Clearly a simple solution of (G.2) is

$$\psi_B(\rho, \theta) = \exp[i(k\rho\cos\theta - \alpha\theta)] \ . \tag{G.3}$$

We shall call this the basic solution ψ_B. On writing

$$x = \rho\cos\theta \ ,$$

and ignoring the term in α, ψ_B represents a plane wave moving along the axis O_x.

In general the phase of ψ_B is

$$\phi \equiv \frac{2\pi x}{\lambda} - \alpha\theta \ , \qquad \left(k = \frac{2\pi}{\lambda}\right) \ . \tag{G.4}$$

For $0 < \alpha < 1$, ignoring the term $\alpha\theta$ in (G.4) is equivalent to making an error δx in the value of the position x, where

$$|\delta x| < \lambda \ .$$

Obviously such a small effect in the electron wave could only be detected by interference.

On using (G.3) it is clear that the change in phase of $\psi_B(\rho\theta)$ on moving from any point on the negative O_x axis to any point on the positive O_x axis depends on whether we pass to the right or the left of the point O. Going first to the right of O and then comparing with the value obtained on going to the left of O always yields a phase difference

$$-2\pi\alpha \ .$$

This the AB effect itself, and the wave function $\psi_B(\rho, \theta)$ is a particularly simple example of the effect.

The question of the single-valuedness of the wave function in the presence of a cylinder of flux has been discussed in Sect. 2.9. The same discussion applies to the line of flux here. The relation between $\psi_B(\rho, \theta)$ and the plane wave $\psi_0(\rho, \theta)$ [in (G.6) below] is similar to the relation between the two wave functions $\psi_A(x)$ and $\psi_0(x)$ that was discussed first in Sect. 2.3.

The Physical Nature of $\psi_B(\rho, \theta)$. Apart from the phase factor $\exp(-i\alpha\theta)$ the wave function $\psi_B(\rho, \theta)$ describes an electron moving with momentum $k\hbar$ along O_x. In Sect. 2.5 [just before (2.15b)] there is a gauge-invariant expression for the current density operator j. By (G.1) and (G.3) we find

$$\psi_B^* j_\rho \psi_B = -\frac{ek\hbar}{m} \cos\theta \psi_B^* \psi_B \ , \qquad \psi_B^* j_\theta \psi_B = \frac{ek\hbar}{m} \sin\theta \psi_B^* \psi_B \ . \tag{G.5}$$

(Notice that in j_θ a term in ρ^{-1} has cancelled out.)

Therefore

$$\psi_B^* j \psi_B = -\frac{ek\hbar}{m} i_x |\psi_B|^2 \ , \tag{G.5a}$$

where i_x is a unit vector along O_x. This corresponds to unit flow of an electron along i_x (with electron charge $-e$).

It should be emphasised that $\psi_B(\rho, \theta)$ is an exact solution of (G.2); it is not just the incident wave, it is itself a whole solution.

The Aharonov–Bohm Solution $\psi_{AB}^{(P)}(\rho, \theta)$. The solution $\psi_{AB}^{(P)}(\rho, \theta)$ was given first by Aharonov and Bohm [G.1] and its was discussed in considerable detail by Olariu and Popescu [G.2], Peshkin, Talmi and Tassie [G.3], among others.

$\psi_B(\rho, \theta)$ is the incident wave in $\psi_{AB}^{(P)}(\rho, \theta)$, but another solution of (G.2) is added to ψ_B so that $\psi_{AB}^{(P)}(\rho, \theta)$ vanishes as $\rho \to 0$ (except for the cases that

$\alpha = 0$ or $\alpha =$ integer). This differs from $\psi_B(\rho, \theta)$ which has the limiting form $\exp(-i\alpha\theta)$ as $\rho \to 0$.

The infinite series of terms [in (G.11b) below] that yields $\psi_{AB}^{(P)}(\rho, \theta)$ is not uniformly convergent with respect to α in $0 < \alpha < 1$ as $\alpha \to 0$.

For $\alpha = 0$ the solution in (G.3) becomes

$$\psi_0(\rho, \theta) = \exp(ik\rho\cos\theta) \ . \tag{G.6}$$

Also for $\alpha = 0$ the complex conjugate $\psi_0^*(\rho, \theta)$ is a solution. The boundary condition that $\psi(\rho, \theta)$ vanishes at $\rho = 0$ requires the use of the form

$$\psi_1(\rho, \theta) = \sin(k\rho\cos\theta) \ .$$

In the series expression (G.11b) below, for $\psi_{AB}^{(P)}(\rho, \theta)$, each term vanishes as $\rho \to 0$. In $0 < \alpha < \frac{1}{2}$ the leading term behaves like $(k\rho)^\alpha$ as $\rho \to 0$. This is different from $|\psi_1(\rho, \theta)|$ which vanishes as an integer power of $k\rho$ as $\rho \to 0$.

For α non-integral, Aharonov and Bohm [G.1] assumed that $B = 0$ in $R < \rho < \infty$ where R is a small radius. In $0 \leq \rho \leq R$ they have uniform B. They match the solutions on $\rho = R$, and find that as $R \to 0$, only Bessel functions $J_v(kR)$ of positive order v remain. There is zero probability (per unit volume) of finding the electron at $\rho = 0$ in this type of solution.

In (G.3), (G.5a) and (G.6) the incident ray moves along $+O_x$. For large values of $k\rho$, and angles not close to the forward direction, i.e. for

$$(2k\rho)^{1/2} \sin(\theta/2) \gg 1 \ , \tag{G.7}$$

the asymptotic form of $\psi_{AB}^{(P)}(\rho, \theta)$ is (cf. Refs. [G.1], [G.2])

$$\psi_B(\rho, \theta) + \sin\pi\alpha \frac{e^{-i\theta/2}}{\sin(\theta/2)} \frac{e^{i(k\rho - \pi/4)}}{(2\pi k\rho)^{1/2}} \ . \tag{G.7a}$$

Here $\psi_B(\rho, \theta)$ is the incident wave as in (G.3). Notice that in the case of three-dimensional elastic scattering of a plane wave by an isotropic centre of force, the elastically scattered wave [in spherical polar coordinates (r, θ, ϕ)] is of the form

$$\psi = \exp(ikr\cos\theta) + r^{-1}\exp(ikr)f(\theta) \ ,$$

where $kr \gg 1$. The scattering amplitude $f(\theta)$ is a function of θ. (See for example Eq. (36) Chap. I, § 7 in Ref. [G.4].)

Aharonov and Bohm [G.1] concluded that the line of flux along O_z causes a scattering cross-section of the order of

$$d\sigma \simeq \frac{\lambda}{4\pi^2} \frac{\sin^2(\pi\alpha)}{\sin^2(\theta/2)} d\theta \ , \tag{G.7b}$$

where $\lambda = 2\pi/k$ is the electron wavelength. $d\sigma$ is the cross-section (into the angle $d\theta$) per unit of length along O_z. (Further comments on two-dimensional scattering are made below.) Other derivations of this scattering are given in [G.2], [G.5]–[G.7].

The relation of the scattering of spin $1/2$ particles to the Aharonov–Casher effect (Sect. 7.2) is discussed by Hagen [G.8]. The exact solution of the Dirac equation for a charged particle in the presence of a Coulomb field and a Dirac monopole is given in [G.9].

Comments. Both $\psi_B(\rho,\theta)$ and $\psi_{AB}^{(P)}(\rho,\theta)$ are solutions of (G.2). They differ in the boundary conditions that are applied (i. e. $\psi_B(\rho,\theta) \to \exp(-i\alpha\theta)$, whereas $\psi_{AB}^{(P)}(\rho,\theta) \to 0$, as $\rho \to 0$). One solution or the other may be preferable for particular practical measurements. The use of $\psi_{AB}^{(P)}(\rho,\theta)$ thus implies that there is the extra restriction as $\rho \to 0$. This restriction can be regarded as giving rise to the cross-section in (G.7b) above.

There does not appear to be any physical reason why an infinite line of flux (at $\rho = 0$) has to be accompanied by an infinitely high potential wall (of zero radius) that keeps the electrons out of the line $\rho = 0$.

Therefore there is no physical reason to exclude $\psi_B(\rho,\theta)$ of (G.3), which is the simplest solution of (G.2). This solution contains the AB effect but it does not show Aharonov–Bohm scattering.

Mathematical Features. $\psi_0(\rho,\theta)$ of (G.6) is a solution of (G.2) in the case $\alpha = 0$. For real θ and any z there is a convergent expansion

$$\exp(iz\cos\theta) = \sum_{m=-\infty}^{\infty} J_m(z)i^m \exp(im\theta) \ , \tag{G.8}$$

where $J_m(z)$ is a Bessel function of integral order m, and

$$J_{-m}(z) = (-1)^m J_m(z) \ , \qquad (m = \text{integer}) \ . \tag{G.8a}$$

Each term on the right of (G.8) is a solution of (G.2) for the case $\alpha = 0$, on putting $z = k\rho$.

For any *real* number v, the Bessel function $J_v(z)$ obeys

$$\frac{d^2 J_v}{dz^2} + \frac{1}{z}\frac{dJ_v}{dz} + \left(1 - \frac{v^2}{z^2}\right)J_v = 0 \ . \tag{G.9}$$

When v is not an integer, Eq. (G.8a) does not hold, and $J_v(z)$ and $J_{-v}(z)$ are independent functions, both obeying (G.9). Except when v is a negative integer, we have

$$J_v(z) \simeq \frac{z^v}{2^v \Gamma(v+1)} \qquad \text{for } z \simeq 0 \ . \tag{G.9a}$$

The second solution of (G.9) in the case $v = m$, where m is an integer, is $Y_m(z)$, which contains a term $\ln z$. This type of solution is not used here.

Among the solutions of (G.9), only $J_0(z)$ has the property

$$J_0(z) \to 1 \qquad \text{as } z \to 0 \ . \tag{G.10}$$

Two Types of Solution. There are two simple types of solution of (G.2) for the case where α is *not* an integer. For the first type we write

$$\psi = e^{im\theta} f(\rho) \ , \qquad (m = \text{integer}) \ . \tag{G.11}$$

Then $f(\rho)$ obeys (G.9) with $v = m + \alpha$ and $z = k\rho$.
The general solution is

$$f(\rho) = a_m J_{m+\alpha}(k\rho) + b_m J_{-m-\alpha}(k\rho) \ , \tag{G.11a}$$

where a_m, b_m are arbitrary constants. Aharonov and Bohm [G.1] in their scattering solution, use the term whose index is $|m + \alpha|$, namely (for $\alpha \neq$ integer)

$$\psi_{AB}^{(P)}(\rho, \theta) = \sum_{m=-\infty}^{\infty} \exp(i\pi|m + \alpha|/2) J_{|m+\alpha|}(k\rho) \exp(im\theta) \ . \tag{G.11b}$$

The second type of solution of (G.2) is of the form

$$\psi = e^{-i\alpha\theta} e^{im\theta} F(\rho) \ . \tag{G.12}$$

The general solution (on excluding $Y_m(k\rho)$) is

$$F(\rho) = J_m(k\rho) \ , \tag{G.12a}$$

where m is an integer.

Using the expansion in (G.8), a sum of such solutions (as in (G.12)) yields $\psi_B(\rho, \theta)$ of (G.3). It is necessary, in the case where α is not zero or an integer, to take the type of solution in (G.12) (including the term $m = 0$) if we wish to obey the condition

$$|\psi| \to 1 \qquad \text{as} \ \ \rho \to 0 \ , \qquad (\alpha \neq 0) \ .$$

Criticism of AB Scattering. In two-dimensional scattering the scattering amplitude (for large ρ) has to vary as $\rho^{-1/2}$, as is seen in (G.7a). Also, $d\sigma$ has to have the dimension of length.

In the case of scattering by a flux wire having zero radius, the only length available is the wavelength λ, as indeed appears in (G.7b). If the scattering were due to a potential cylinder having $\rho = R$, then R, or $R + \lambda$, would give the scale of the cross-section $d\sigma$.

Although there seems to be no objection in principle to having a fibre of zero width carrying the flux Φ, that is not possible in practice. Most real fibres or whiskers will have $R \gg \lambda$ (where λ is the wavelength of the electron). In such a case the scattering by the material of the fibre will make it impossible to detect $d\sigma$ of (G.7b).

There would presumably be a difficulty *in principle* concerning AB scattering, if it was thought to be due to some interaction that extended in space away from the fibre. The induction B is zero outside the line O_z, and no other force is present in space that could cause scattering.

However there can be an effect arising at the surface of the fibre; an effect which remains as the radius goes to zero. In discussing the original AB effect in Sect. 5.3, we used the boundary condition on the electron wave function on going once around the cylinder. This boundary condition remains valid as the radius of the cylinder goes to zero, and it could give rise to the AB scattering. The effect of the boundary condition could extend outwards a distance of the order λ away from the axis O_z. (The behaviour of the electron wave function close to the fibre is not simple: it is discussed in Ref. [G.2].)

The AB scattering is purely a quantum effect which will disappear as $\lambda \to 0$, (on keeping the other parameters – velocity, distance, and angles, etc. – constant). Thus AB scattering is not entirely to be subjected to criticisms that are based on classical concepts.

Comment on Scattering in Two Dimensions. Ideas on scattering that have arisen from studying three-dimensional problems need not be valid for the pure two-dimensional situation.

It is sufficient to compare solutions of Laplace's equation to see the situation. For differentiable functions $\phi(x)$ and $\psi(x)$ we use the identity

$$\int_s dS \left(\phi \frac{\partial \psi}{\partial n} - \psi \frac{\partial \phi}{\partial n} \right) = \int_V d\tau \left(\phi \nabla^2 \psi - \psi \nabla^2 \phi \right) \ . \tag{G.13}$$

The integral on the right is over a volume V and that on the left is over the surface S, or all the surfaces S, of this volume. $\partial/\partial n$ denotes the derivative along the *outward* normal n.

Consider the case of three-dimensions: We are interested in the value $\psi(P)$ and ψ at a point P expressed in terms of the value of $\nabla^2 \psi$ throughout V plus the values of ψ and $\partial \psi/\partial n$ on a boundary surface S. This is done by using $\phi = 1/r(P)$ where $r(P)$ is the distance from P. Equation (G.13) yields

$$4\pi \psi(P) = - \int_V d\tau \frac{1}{r} \nabla^2 \psi + \text{(surface terms)} \ . \tag{G.14}$$

If it is assumed that $|\psi|$ decreases as r^{-1} (or faster) for large r, the surface terms will not give us trouble here. Also it is clear that the "source term"

$$(d\tau) \nabla^2 \psi \ ,$$

assuming it is of given strength, becomes much less important if it is moved further away from P so that r^{-1} decreases. This feature supports the naive idea of scattering in the three-dimensional case. Far away parts of space do not trouble us unduly.

In the case of two-dimensions the corresponding form of ϕ is

$$\phi = \ln \rho(P) \ ,$$

where $\rho(P)$ is the distance from the line of observation (P) to the current (line) position. Analogous to (G.14) we have

$$2\pi \psi(P) = + \int_V d\tau \nabla^2 \psi \cdot \ln \rho(P) + \text{(surface terms)} \ . \tag{G.15}$$

Here all quantities are independent of the coordinate z, and we think of a slice of unit height in z.

The importance of the source term $(\mathrm{d}\tau)\nabla^2\psi$ will in this case increase as $\rho(P)$ is increased (other things being equal). This indicates a basic difference between the two-dimensional and three-dimensional cases.

Finally notice that if the two-dimensional source is replaced by a source at the same value of (ρ, θ) but extending from $z = -L$ to $z = +L$ (instead of from $z = -\infty$ to $z = +\infty$) then, *as a three-dimensional problem*, we have in place of r^{-1} the term

$$\int_{z=-L}^{z=+L} \frac{\mathrm{d}z}{r} = 2(-\ln \rho(P) + \text{const.}) \ . \tag{G.16}$$

The term "const." here varies with L, and when it is absorbed into one of the surface terms, this relation (G.16) shows the connection between the terms containing $\nabla^2\psi$ in (G.14) and (G.15).

The origin of $\ln \rho(P)$ in (G.15) is thus no mystery, but its presence shows that two-dimensional problems may not have simple intuitive interpretations for those accustomed to working in three dimensions.

Appendix H. Quantum Theory
of Cyclotron Motion

This appendix gives a brief account of the quantum methods available to describe the motion of an electron in a uniform induction field (in addition to the quantum minimisation method used in Sects. 3.1–3.3). These are first, the method commonly called Landau–Lifschitz, where gauge invariance is to be used to give a meaningful solution. This method also shows the uncertainty of the centre of the orbit, in agreement with the basic quantum uncertainty in flux, as given in Sect. 6.4. Secondly in a more symmetric gauge, the Schrödinger equation solutions are closely related to the rotational solutions for a two-dimensional harmonic oscillator. Certain solutions of the latter in turn are expressible as sums of bilinear terms of single oscillator wave functions. These considerations enable us to study the solutions, and in particular, the radial motion of the electron in the uniform induction. Moreover, the gauge invariant nature of the current operator becomes clear.

There are several ways of determining the motion of an electron having charge $-e$ in a uniform induction field B. In Sects. 3.1–3.3 a solution is given that depends notably on the gauge chosen [(3.1)], and uses the principle of minimum expectation values. This solution gives a simple physical description of the motion.

A well known solution is that given by Landau and Lifschitz (pp. 456–458 of Ref. [H.1], or pp. 313–315 of Ref. [H.2]. See also Landau [H.3]). This differs from our solution in Sects. 3.1–3.3 in several ways. It uses a different gauge and it is purely wave mechanical. However it does not contain the radial oscillations of Sect. 3.3. We shall examine the physical interpretation of Landau's solution.

This solution is easiest to envisage in terms of the expectation values for quantum operators of position, velocity, etc. Then the electron's position is seen to move round a circle in any plane orthogonal to B. The centre of this circle is uncertain by an area that contains one quantum unit of flux $\phi_e = h/e$. This is in agreement with the accuracy of flux measurement by a single electron, that was described in Chap. 5.

A complete wave mechanical solution for motion perpendicular to the uniform induction B is based on the solution of the two-dimensional oscillator problem that was treated in Sect. 8.9. This solution makes considerable use of generalised Laguerre polynomials. The wave mechanical solution that we can then use, describes the motion around a circle and the spread due to radial vibrations in one formula. The wave function has the interesting feature that it is easier to describe in terms of ρ^2 than ρ itself (ρ is the radial distance). Early results were presented by Page in [H.4].

H.1 Landau–Lifschitz Solution

We use the gauge

$$\mathbf{A} = (-By, 0, 0) \tag{H.1}$$

for the motion of an electron in the uniform induction $\mathbf{B}$ along O_z. Schrödinger's equation is then

$$H\psi \equiv \frac{1}{2m} \left[(p_x - eBy)^2 + (p_y^2 + p_z^2) \right] \psi = E\psi \ . \tag{H.2}$$

The momentum operators p_x and p_z commute with H itself, and they can be taken as constants of motion, p_x' and p_z' respectively. Thus there is a solution of the form

$$\psi_{LL} = \exp\left[(i/\hbar)(p_x' x + p_z' z) \right] \chi(y) \ , \tag{H.2a}$$

and $\chi(y)$ obeys

$$-\frac{\hbar^2}{2m} \frac{d^2\chi}{dy^2} + \frac{(eB)^2}{2m} (y - y_0)^2 \chi = \left(E - \frac{p_z'^2}{2m} \right) \chi \tag{H.3}$$

with

$$y_0 = p_x'/eB \ . \tag{H.3a}$$

This is sometimes regarded as the required answer.

In the Oxz plane the electron's motion appears to be described by a plane wave, while (H.3) describes simple harmonic oscillations in the direction O_y with circular frequency

$$eB/m - \omega_B \ . \tag{H.3b}$$

The wave function

$$\chi(y) = \exp\left[\frac{-eB}{2\hbar} (y - y_0)^2 \right] H_n\left(\left(\frac{eB}{\hbar} \right)^{1/2} (y - y_0) \right) \tag{H.4}$$

describes this motion along O_y. H_n with $n = 0, 1, 2, \ldots$ are Hermite polynomials and the eigenvalues of energy are

$$E_n = \frac{p_z'^2}{2m} + \left(n + \frac{1}{2} \right) \hbar \omega_B \ . \tag{H.4a}$$

Any value of p_z' is allowed.

Physical Interpretation. Care is needed on account of the gauge. For example a constant g may be added to $(-By)$ in (H.1). This will alter y_0 and the "centre" of the wave function $\chi(y)$ of (H.4) will thereby be moved. Moreover it cannot be the case that the motion in the plane Oxz is rectilinear; that cannot be the correct interpretation.

In the proper interpretation, only gauge-invariant quantities are to be used. The gauge-invariant form of the current density is given in Sect. 2.5 as

$$j = -\frac{\hbar e}{2mi}(\psi^*\boldsymbol{\nabla}\psi - \psi\boldsymbol{\nabla}\psi^*) - \frac{e^2}{m}\psi^*\boldsymbol{A}\psi \ . \tag{H.5}$$

For convenience we shall use the velocity operator

$$v = -j/e \ .$$

We put $p_z' = 0$, so the motion is two dimensional, and we consider matrix elements between states of the form

$$\exp(i\boldsymbol{p}\cdot\boldsymbol{x}/\hbar) \ .$$

Equation (H.5) then yields the operator relation

$$v_x = p_x/m - \omega_B y \ . \tag{H.6}$$

In place of (H.3a) we have

$$v_x = \omega_B(y_0 - y) \ , \tag{H.6a}$$

where y_0 is a constant of motion.

The operator relation

$$[(p_y/eB) - x, p_x - eBy] = 0$$

shows that

$$(p_y/eB) - x$$

commutes with H of (H.2). Thus

$$p_y/eB = x - x_0 \tag{H.7}$$

where x_0 is a constant of motion. It follows from (H.5) and (H.7) that the other (the y-)component of the velocity operator is

$$v_y = \omega_B(x - x_0) \ . \tag{H.7a}$$

Equations (H.6a) and (H.7a) are gauge invariant.

The quantities x_0 and y_0 are constants of motion but they do not mutually commute. On using the definition of the operator y_0 in (H.6a), we have the commutation relation:

$$eB[x_0, y_0] = [x - (p_y/eB), p_x] = i\hbar \ . \tag{H.8}$$

Therefore x_0 and y_0 cannot be diagonal for the same state vector. They cannot both have sharp values in the same state.

The Uncertainty of the Centre. Equations (H.6a) and (H.7a) show that the expectation values of the vectors

$$(v_x, v_y) \qquad \text{and} \qquad (x - x_0, y - y_0)$$

are always orthogonal. Thus the electron moves with angular frequency ω_B around a circle whose centre is (x_0, y_0). The sense of the rotation is positive about O_z, as is expected from the Lorentz force in (2.1) in the case $q = -e$.

This picture is not quite correct because of the quantum effect in (H.8). In Sect. 6.1 we gave various measures of the quantum uncertainty. In particular Δ denoted the *practical definition*, and it is seen from (6.5) that for a canonical conjugate pair of operators p, q, this gives

$$\Delta p \Delta q \geq h \ .$$

It follows from (H.8) that

$$eB \Delta x_0 \Delta y_0 \geq h \ . \tag{H.9}$$

The classical picture of the electron moving on a circle is thus incomplete, in that the centre of the circle is blurred somewhat, and its position cannot be stated precisely. The amount of blurring corresponds to an uncertainty in flux of magnitude

$$\Delta \Phi = B \Delta x_0 \Delta y_0 \geq h/e = \phi_e \ .$$

This uncertainty agrees with the basic quantum uncertainty in measuring flux by using one charge particle, as is discussed in Sect. 6.4.

H.2 The Symmetric Schrödinger Equation Solution

In Eq. (3.1) the more symmetric gauge

$$\boldsymbol{A} = (0, 0, \rho B/2) \tag{H.10}$$

is introduced. This describes uniform induction $\boldsymbol{B}$ along O_z.

Imagine an electron in a two-dimensional harmonic oscillator in the plane Oxy. The pure oscillations have the circular frequency Ω. The electron is also under the influence of $\boldsymbol{B}$. As in (8.39), Schrödinger's equation, in cylindrical polar coordinates (z, ρ, θ) becomes

$$i\hbar \frac{\partial \psi}{\partial t} = \left\{ -\frac{\hbar^2}{2m} \left[\frac{\partial^2}{\partial z^2} + \frac{1}{\rho} \frac{\partial}{\partial \rho} \left(\rho \frac{\partial}{\partial \rho} \right) \right] \right.$$
$$\left. + \frac{1}{2m} \left(\frac{\hbar}{i\rho} \frac{\partial}{\partial \theta} + \frac{e\rho B}{2} \right)^2 + \frac{m\Omega^2 \rho^2}{2} \right\} \psi \ . \tag{H.11}$$

We will use soltions ψ of (H.11) that contain the factor

$$\exp(\pm il\theta - iEt/\hbar) \ , \qquad (l = 0, 1, 2, \ldots) \ . \tag{H.11a}$$

Then $l\hbar$ is the angular momentum about O_z. There will be no other dependence on θ and t. E is the energy.

$\psi_{\Omega l}$ and ψ_{Bl} will denote solutions ψ that apply to the cases $B = 0$ and $\Omega = 0$, respectively. These solutions have the same boundary conditions for $|\psi|$. The corresponding values of the energy E are denoted $E_{\Omega l}$ and E_{Bl}.

On substituting

$$\Omega = \omega_B/2 \quad , \qquad E_{Bl} = E_{\Omega l} + l\hbar\Omega \quad , \tag{H.12}$$

the solution $\psi_{\Omega l}$ becomes the solution ψ_{Bl}, as can be seen from (H.11).

The method used here is a special case of the phase γ that was used in Sect. 8.10. The term in (H.11) that is linear in B is the anholonomic term that gives the phase γ as in (8.59). The case here of the uniform field B is simple since $n = 1$. The energy $l\hbar\Omega$ in (H.12) is just E_{AN} as given in (8.69b). The Hamiltonian H_0 of (8.58) is just the oscillator Hamiltonian.

Oscillator Solutions. The two-dimensional *oscillator* solutions $\psi_{\Omega l}$ are obtained by writing

$$\psi_{\Omega l} = \rho^l G(\rho) \quad , \qquad l = 0, 1, 2, \ldots \quad .$$

By virtue of (H.11), $G(\rho)$ obeys

$$\frac{d^2 G}{d\rho^2} + \frac{2l+1}{\rho}\frac{dG}{d\rho} + \left(K^2 - \frac{\rho^2}{a^4}\right)G = 0 \quad , \tag{H.13}$$

where

$$a^2 = \hbar/m\Omega \quad , \qquad K^2 = 2mE_{\Omega 1}/\hbar^2 \quad , \tag{H.13a}$$

and a and K^{-1} are positive constants having the dimensions of length. $G(\rho)$ has to be finite as $\rho \to 0$, and it must vanish faster than ρ^{-1} as $\rho \to \infty$.

The final substitutions,

$$y = \rho^2/a^2 \quad , \qquad G(\rho) = F(y)\exp(-y/2)$$

yield

$$yF'' + (l+1-y)F' + \frac{1}{2}\left(\frac{E_{\Omega l}}{\hbar\Omega} - l - 1\right)F = 0 \quad . \tag{H.14}$$

Here the prime denotes differentiation with respect to y.

The condition that $F(y)$ should be a polynomial in y of finite order is necessary, and it yields the eigenvalues of energy

$$E_{\Omega l} = \hbar\Omega(l + 1 + 2n) \quad , \tag{H.15}$$

where $n = 0, 1, 2, 3, \ldots$. Notice that the lowest, or zero-point, eenergy is $E_{\Omega l} = \hbar\Omega$. It has this value because there are zero-point oscillations independently in each dimension O_x and O_y.

The corresponding eigenstate wave functions are

$$\psi_{\Omega l} = \text{const.} \times \rho^l \exp\left(\frac{-\rho^2}{2a^2}\right) L_n^{(l)}\left(\frac{\rho^2}{a^2}\right) e^{\pm il\theta} e^{-iE_{\Omega l}t/\hbar} \quad . \tag{H.16}$$

The scale of length a is given by (H.13a). $L_n^{(l)}(\beta)$ are generalised Laguerre polynomials [H.5], [H.6]. These orthogonal polynomials of degree n have the form

$$L_n^{(\alpha)}(\beta) = n! \sum_{m=0}^{n} (-1)^m \left\{ \begin{matrix} n+\alpha \\ n-m \end{matrix} \right\} \frac{\beta^m}{m!} \quad , \tag{H.17}$$

where α is any real number in $\alpha > -1$. The standardisation used in (H.17) is

$$(-1)^n \beta^n \quad .$$

The general solution $\psi_{\Omega l}$ is a linear combination of the eigenstates in (H.16).

The polynomial $L_n^{(\alpha)}(\beta)$ has n zeros in $0 < \beta < \infty$, and these separate the $n+1$ zeros of $L_{n+1}^{(\alpha)}(\beta)$. This is true for $n = 1, 2, \ldots$. As simple examples we have

$$L_0^{(\alpha)}(\beta) = 1 \quad , \qquad L_1^{(\alpha)}(\beta) = \alpha + 1 - \beta \quad ,$$
$$L_2^{(\alpha)}(\beta) = (\alpha + 2)(\alpha + 1) - 2(\alpha + 2)\beta + \beta^2 \quad ,$$

and so on.

Normalisation of $\psi_{\Omega l}$ depends on the relation

$$\int_0^\infty d\beta e^{-\beta} \beta^\alpha L_n^{(\alpha)}(\beta) L_{n'}^{(\alpha)}(\beta) = n!\, \Gamma(n + \alpha + 1)\delta_{nn'} \quad . \tag{H.17a}$$

Another useful relation is

$$\int_0^\infty d\beta e^{-\beta} \beta^{\alpha+1} \left(L_n^{(\alpha)}(\beta) \right)^2 = (2n + \alpha + 1)n!\, \Gamma(n + \alpha + 1) \quad . \tag{H.17b}$$

Expectation Values of ρ^2. Equations (H.17a) and (H.17b) show that for the state $\psi_{\Omega l}$

$$\langle \rho^2 \rangle = \frac{\int_0^\infty \rho^2 |\psi_{\Omega l}|^2 \rho\, d\rho}{\int_0^\infty |\psi_{\Omega l}|^2 \rho\, d\rho} = (2n + l + 1)a^2 \quad . \tag{H.18}$$

This result is related to that in (H.15). Applying the principle of equipartition of energy to the Hamiltonian in (H.11) in the case $B = 0$, gives directly

$$E_{\Omega l} = m\Omega^2 \langle \rho^2 \rangle; \tag{H.19}$$

and this result also follows from (H.15) and (H.18).

Expansion in Terms of One-Dimensional Oscillator Solutions. In the case of the two-dimensional harmonic oscillator (i. e. the case $B = 0$), the Hamiltonian (H.11) is separable into an operator in x plus an operator in y. Each of these relates to simple harmonic motion. The eigenstates are

$$\psi_n(x) = \exp(-x^2/2a^2) H_n(x/a) \quad , \tag{H.20}$$

and a similar expression for $\psi_n(y)$. The length a is given by (H.13a). The H_n are Hermite polynomials, such as occur in (H.4).

Thus there are solutions of (H.11) of the form

$$\psi_S = \psi_{n'}(x)\psi_{n''}(y) \quad . \tag{H.20a}$$

The energy in state ψ_S is

$$E_S = (n' + n'' + 1)\hbar\Omega \quad , \qquad (n', n'' = 0, 1, 2, \ldots) \quad . \tag{H.20b}$$

Using $a = 1$ for convenience, we can write the time-independent factors in $\psi_{\Omega l}$ of (H.16) as

$$\exp(-\rho^2/2)L_n^{(l)}(\rho^2)(x \pm iy)^l \quad . \tag{H.20c}$$

On choosing n', n'' so that

$$n' + n'' = 2n + 1 \tag{H.20d}$$

we ensure that ψ_S of (H.20a) is degenerate with $\psi_{\Omega l}$.

This suggests that there is an expansion of the form:

$$\rho^l e^{il\theta} L_n^{(l)} = \sum_{n',n''} b_{n'n''} H_{n'}(x) H_{n''}(y) \quad , \tag{H.20e}$$

where n', n'' are zero or positive integers satisfying (H.20d). Also

$$|b_{n'n''}| = |b_{n''n'}| \quad .$$

We have not proved this statement, but merely give a few examples:
$l = 1$, $n = 0$:

$$2\rho e^{i\theta} L_0^{(1)} = H_1(x)H_0(y) + iH_0(x)H_1(y) \quad .$$

$l = 0$, $n = 1$:

$$-4L_1^{(0)}(\rho^2) = H_2(x)H_0(y) + H_0(x)H_2(y) \quad .$$

$l = 1$, $n = 1$:

$$-8\rho e^{i\theta} L_1^{(1)}(\rho^2) = iH_0(x)H_3(y) + H_3(x)H_0(y)$$
$$+ H_1(x)H_2(y) + iH_2(x)H_1(y) \quad .$$

$l = 3$, $n = 0$:

$$8\rho^3 e^{3i\theta} L_0^{(3)}(\rho^2) = -iH_0(x)H_3(y) + H_3(x)H_0(y)$$
$$- 3H_1(x)H_2(y) + 3iH_2(x)H_1(y) \quad .$$

The Energy Degeneracy. These wave functions provide examples of how the degeneracy in energy in the case of two simple oscillators allows us to build up general rotational states like $\psi_{\Omega l}$ of (H.16).

What happens in the case that $\Omega = 0$ but $B \neq 0$? On restricting ourselves to solutions in which the dependence on θ and t is of the form in (H.11a), we go from solutions of equations for $B = 0$, $\Omega \neq 0$ to solutions of equations for $B \neq 0$, $\Omega = 0$, on using (H.12) and (H.16). Thus equations similar to (H.20e) will again apply. In other words, *when* the condition in (H.11a) is applied, the operator on the right of (H.11) is separable, even for $B \neq 0$. At this point it is useful to look at the remarks in Sect. 8.10 in connection with the phase γ.

H.3 Motion in Uniform Induction B

In case $\Omega = 0$, $B \neq 0$, Eq. (H.11) yields the solution ψ_{Bl} which is obtained from (H.16) by the substitutions in (H.12). Thus

$$\psi_{Bl} = \text{const.} \times \rho^l \exp\left(\frac{-\rho^2}{2a'^2}\right) L_n^{(l)}\left(\frac{\rho^2}{a'^2}\right) e^{\pm il\theta} e^{-iE_{Bl}t/\hbar} \tag{H.16a}$$

and the energy is, in place of (H.15),

$$E_{Bl} = \left(n + l + \tfrac{1}{2}\right)\hbar\omega_B \; . \tag{H.21}$$

The length scale now is a' defined by

$$a'^2 = \frac{2\hbar}{eB} = \frac{\phi_e}{\pi B} \; . \tag{H.22}$$

A circle of radius a' in the Oxy plane encloses one quantum unit of flux ϕ_e. Equation (H.18) yields

$$\frac{eB}{2}\langle\rho^2\rangle = (2n + l + 1)\hbar \tag{H.23}$$

for the expectation value of ρ^2 in state ψ_{Bl}.

Equipartition Principle. In the oscillator case [i.e. $B = 0$ in (H.11)], Eqs. (H.15) and (H.18) show that the ratio

$$E_{\Omega l}/\langle\rho^2\rangle$$

is independent of the quantum numbers. Equations (H.21) and (H.23) show that this is no longer true for motion in uniform induction B. Such motion is not invariant under time reversal, and the equipartition principle will not be meaningful.

Equations (H.21) and (3.10) show that the method used in Sects. 3.1–3.3 gives the same eigenvalues of energy. The quantum numbers are related by

$$M \to l \; , \qquad n \to n \; . \tag{H.23a}$$

The radius ρ_0 of the orbit in Sect. 3.1 is related to (H.23) by (3.4), which again gives $M \to l$ (assuming that in this case $l \gg 2n + 1$).

Large Value of $\langle\rho^2\rangle$. The approximation that was used in Sects. 3.1 and 3.3 for the Hamiltonian term

$$\frac{1}{2m}\left(\frac{l\hbar}{\rho} + \frac{eB\rho}{2}\right)^2 \tag{H.24}$$

is

$$\frac{(eB)^2}{2m}\left[\rho_0^2 + (\rho - \rho_0)^2\right] \; . \tag{H.24a}$$

This is only accurate for small $|\rho - \rho_0|$, and for quantum number n it yields [cf. (3.9c)]

$$\langle(\rho - \rho_0)^2\rangle = \frac{\hbar}{eB}\left(n + \frac{1}{2}\right) \; . $$

In fact the term in (H.24) is

$$\frac{1}{2m}\frac{l^2\hbar^2}{\rho^2} + \frac{1}{2}l\hbar\omega_B + \frac{1}{8}m\omega_B^2\rho^2 \ . \tag{H.24b}$$

The second and third terms here give a slowly rising function of ρ^2 and this in turn causes a much larger value of $\langle\rho^2\rangle$ in (H.23) than would be suggested by (3.9c).

The physical properties involved here are:

(a) that the uniform induction B provides a much softer restoring force, acting after any outward deviation, than would be given by the linear restoring force used in (3.9);

(b) that (H.23) relates to two dimensions, whereas (3.9c) is one dimensional.

The Radial Structure. The radial functions $L_n^{(l)}(\rho^2/a'^2)$ in ψ_{Bl} of (H.16a) will give the structure of any oscillatory radial motion about the basic cyclotron solution having $n = 0$. That solution of course has a "zero-point width".

The mathematical manipulations of those solutions ψ_{Bl} are easier on using the variable ρ^2 than they would be on using ρ. Thus we have not quoted the value of $\langle\rho\rangle$. It is easy, in addition to (H.18), to evaluate

$$\langle(\rho^4) - \langle\rho^2\rangle^2\rangle = [n(n + 1) + (n + 1)(n + l + 1)]\, a'^4 \ . \tag{H.25}$$

This gives some idea of the role of the spread of the orbit in the radial direction, associated with the quantum number n.

If we use a simple model in which the values of ρ

$$\rho_0 - \Delta \ , \qquad \rho_0 \ , \qquad \rho_0 + \Delta \ ,$$

occur with the weights 1/4, 1/2, 1/4 respectively, and we assume that Δ/ρ_0 is small, then

$$\langle\rho^2\rangle = \rho_0^2 + \Delta^2/2 \ ,$$
$$\langle\rho^4\rangle = \rho_0^4 + 3\rho_0^2\Delta^2 \ .$$

We therefore estimate that

$$\Delta^2 \simeq \frac{\langle\rho^4\rangle - (\langle\rho^2\rangle)^2}{2\langle\rho^2\rangle} \ .$$

Now using (H.23) and (H.25) yields

$$\Delta^2 = a'^2 \frac{n(n + l) + (n + 1)(n + l + 1)}{2(2n + l + 1)} \ . \tag{H.25a}$$

For the case that the radial spread of the motion is not very large we expect that

$$l \gg n \ . \tag{H.25b}$$

In this case (H.25a) becomes

$$\Delta^2 \simeq a'^2 \left(n + \tfrac{1}{2}\right) \tag{H.25c}$$

and 2Δ gives an estimate of the radial spread of ψ_{Bl}. The variance Δ_v about ρ_0 in our simple model is given by

$$\Delta_v^2 = \Delta^2/2 \ .$$

By (H.22) and (H.25b) we have

$$\Delta_v^2 \simeq \left(n + \frac{1}{2}\right) \frac{\phi_e}{2\pi B} \ . \tag{H.26}$$

This is just the value of

$$\langle (\rho - \rho_0)^2 \rangle_{(n)}$$

given in (3.9c). The suggested validity condition ($n \ll 4M$) involved in (3.9c) is roughly the same as (H.25b) above.

Gauge Invariance. How does gauge invariance, as given in (2.16), apply to the solution ψ_{Bl} for motion in uniform induction $\boldsymbol{B}$? The simplest way to answer this is to look at the current density operator $\boldsymbol{j}$ of Sect. 2.5 or (H.5) above.

Only the azimuthal component j_θ is of interest. We have

$$j_\theta = -\frac{\hbar e}{mi}\psi^* \frac{1}{\rho}\frac{\partial \psi}{\partial \theta} - \frac{e^2}{m}\psi^* A_\theta \psi \ , \tag{H.27}$$

and, as in (H.10),

$$A_\theta = B\rho/2 \ .$$

For $l \gg n$ and with n not large, the wave function ψ_{Bl} of (H.16a) will only be spread over a small range of values of ρ. If L_z' is the eigenvalue of the operator L_z in (2.13a), then (H.27) gives the eigenvalue

$$j_\theta = -\frac{eL_z'}{m\rho} - \frac{e^2\rho}{2m}B \ . \tag{H.28}$$

Now

$$L_z' = m\rho v/2 \ , \tag{H.28a}$$

where v is the eigenvalue of velocity. The factor $1/2$ in (H.28a) is correct because of the form of the angular momentum term in (3.2a) or as is seen from (H.11) and (H.24) above. [It is important to remember that L_θ is defined by (2.13a) alone in wave mechanics.]

Thus

$$\frac{eL_z'}{m\rho} = \frac{1}{2}ev \ , \qquad \frac{e^2\rho}{2m}B = \frac{e\rho\omega_B}{2} = \frac{ev}{2} \ ,$$

so finally we have the simple gauge-invariant result:

$$j_\theta = -ev \ .$$

In the case of the Landau–Lifschitz solution in Sect. H.1, the use of the current density $\boldsymbol{j}$ also showed the simple gauge-invariant result.

References

Chapter 1

1.1 W. Franz; Verh. Deutsche Phys. Ges. No. **2**, 65 (1939)
1.2 W. Ehrenberg, R. E. Siday: Proc. Phys. Soc. London **62B**, 8 (1949)
1.3 Y. Aharonov, D. Bohm: Phys. Rev. **115**, 485 (1959)
1.4 R. G. Chambers: Phys. Rev. Lett. **5**, 3 (1960)
1.5 S. Kamefuchi (ed): Proceedings of the International Symposium; Foundations of Quantum Theory. Phys. Soc. Japan (1984)
1.6 M. Peshkin, A. Tonomura: *The Aharonov–Bohm Effect*. Lecture Notes in Physics **340** (Springer, Berlin, Heidelberg (1989)
1.7 S. Olariu, I. I. Popescu: Rev. Mod. Phys. **57**, 339 (1985)

Chapter 2

2.1 Y. Aharonov, D. Bohm: Phys. Rev. **115**, 485 (1959)
2.2 W. Ehrenberg, R. E. Siday: Proc. Phys. Soc. London **62B**, 8 (1949)
2.3 L. J. Tassie, M. Peshkin: Ann. Phys. (N.Y.) **16**, 177 (1961)
2.4 T. T. Wu, C. N. Yang: Phys. Rev. **D 12**, 3845 (1975)
2.5 M. Peshkin, A. Tonomura: *The Aharonov–Bohm Effect*. Lecture Notes in Physics **340** (Springer, Berlin, Heidelberg (1989)
2.6 V. F. Weisskopf: *Lectures in Theoretical Physics III*, ed. by W. E. Brittin, B. W. Downs, J. Downs (Interscience, New York 1961) pp. 63–70
2.7 E. Noerdlinger: Nuovo Cimento **23**, 158 (1962)
2.8 N. Peshkin: Phys. Report **80**, No 6, 375 (1981)
2.9 F. Wilczek: Phys. Rev. Lett. **48**, 1144 (1982)
2.10 D. H. Kobe: J. Phys. A **15**, L543 (1982)
2.11 P. Carruthers, N. N. Nieto: Rev. Mod. Phys. **40**, 411 (1968)
2.12 S. M. Barnett, D. T. Pegg: Phys. Rev. **A 41**, 3427 (1990)
2.13 F. Bloch: Z. Physik **52**, 555 (1928)
2.14 A. M. G. Floquet: Ann. de l'Ecole Norm. Sup. (2) XII, 47 (1883)
2.15 C. Kittel: *Quantum Theory of Solids* (2nd edn) (Wiley, New York 1987)
2.16 C. Kittel: *Introduction to Solid State Physics* (2nd edn) (Wiley, New York 1957)
2.17 R. Rajaraman: *Solitons and Instantons* (North Holland, Amsterdam 1982)
2.18 J. Clarke: Nature (London) **352**, 110 (1991)
2.19 A. Baroni, G. Paternò: *Physics and Applications of the Josephson Effect* (Wiley, New York 1982)
2.20 J. Clarke: Nature (London) **372**, 501 (1994)

Chapter 3

3.1 J. M. Ziman: *Principles of the Theory of Solids* (2nd ed.) (Cambridge University Press, Cambridge 1979)
3.2 C. Kittel: *Quantum Theory of Solids* (2nd edn) (Wiley, New York 1987)
3.3 L. D. Landau, E. M. Lifshitz: *Quantum Mechanics, Non-Relativistic Theory* (3rd ed.) (transl. J. B. Sykes and J. S. Bell) (Pergamon, Oxford 1977)
3.4 V. I. Ginsburg, L. D. Landau: Zh. Eksp. Teor. Fiz. **20**, 1064 (1950)
3.5 V. L. Ginsburg, L. D. Landau: Fiz. Tverdoga Tela **2**, 2031 (1960) [Engl. trans.: Sov. Phys. Solid State **2**, 1824 (1960)]
3.6 P. G. de Gennes: *Superconductivity of Metals and Alloys* (Benjamin, New York 1966)
3.7 D. R. Tilley, J. Tilley: *Superfluidity and Superconductivity* (2nd ed.) (Hilger, Bristol 1986)
3.8 A. G. Aronov, Yu. V. Sharvin: Rev. Mod. Phys. **59**, 755 (1987)
3.9 M. S. Livingston: *High Energy Accelerators* (Interscience, New York 1954)

Chapter 4

4.1 J. M. Burgers: Ann. der Physik **52**, 195 (1917); E. Fues: *Handbuch der Physik*, Band V, Grundlagen der Mechanik (Springer, Berlin 1927) pp. 148–150
4.2 V. I. Arnold: *Mathematical Methods of Classical Mechanics* (2nd ed.) (Springer, New York 1989) p. 298 ff.

Chapter 5

5.1 M. Peshkin, A. Tonomura: *The Aharonov–Bohm Effect*. Lecture Notes in Physics **340** (Springer, Berlin, Heidelberg (1989)
5.2 S. Olariu, I. I. Popescu: Rev. Mod. Phys. **57**, 339 (1985)
5.3 G. Möllenstedt, H. Düker: Z. Phys. **145**, 377 (1956)
5.4 H. Boersch, H. Hamisch, K. Grohmann, D. Wohlleben: Z. Phys. **165**, 79 (1961); Z. Phys. **167**, 72 (1962)
5.5 R. G. Chambers: Phys. Rev. Lett. **5**, 3 (1960)
5.6 H. A. Fowler, L. Marton, J. A. Simpson, J. A. Suddeth: J. Appl. Phys. **32**, 1153 (1961)
5.7 G. Möllenstedt, W. Bayh: Naturwissenschaften **49**, 81 (1962)
5.8 W. Bayh: Z. Phys. **169**, 492 (1962)
5.9 S. Olariu, I. I. Popescu: Rev. Mod. Phys. **57**, 339 (1985)
5.10 Y. Aharonov, D. Bohm: Phys. Rev. **123**, 1511 (1961)
5.11 N. G. van Kampen: Phys. Lett. **106A**, 5 (1984)
5.12 T. Troudet: Phys. Lett. **111A**, 274 (1985)
5.13 R. A. Brown, D. Home: Il Nuovo Cimento **107B**, 303 (1992)
5.14 B. R. Holstein: Am. J. Phys. **59**, 1080 (1991)
5.15 Y. Aharonov, D. Bohm: Phys. Rev. **125**, 2192 (1962)
5.16 S. Kamefuchi (ed): Proceedings of the International Symposium; Foundations of Quantum Theory. Phys. Soc. Japan (1984)
5.17 M. Peshkin, A. Tonomura: *The Aharonov–Bohm Effect*. Lecture Notes in Physics **340** (Springer, Berlin, Heidelberg (1989)
5.18 B. S. DeWitt: Phys. Rev. **125**, 2189 (1962)

5.19 F. J. Belinfante: Phys. Rev. **128**, 2832 (1962)
5.20 S. M. Roy: Phys. Rev. Lett. **44**, 111 (1980)
5.21 U. Klein: Phys. Rev. **D 23**, 1463 (1981)
5.22 P. Boccieri, M. Loinger: Nuovo Cimento **47A**, 475 (1978)
5.23 P. Bocchieri, A. Loinger, G. Siragusa: Nuovo Cimento **56A**, 55 (1980)
5.24 Y. Aharonov, D. Bohm: Phys. Rev. **115**, 485 (1959)
5.25 D. M. Greenberger: Phys. Rev. **D 23**, 1460 (1981)
5.26 D. Greenberger, A. W. Overhauzer: Rev. Mod. Phys. **51**, 43 (1979)
5.27 P. Bocchieri, A. Loinger, G. Siragusa: Nuovo Cimento **51A**, 1 (1979)
5.28 P. Bocchieri, A. Loinger: Lett. Nuovo Cimento **30**, 1949 (1981)
5.29 H. Boersch, H. Hamisch, K. Grohmann: Z. Phys. **169**, 263 (1962)
5.30 H. J. Lipkin: Phys. Rev. **D 23**, 1466 (1981)
5.31 H. Erlichson: Am. J. Phys. **38**, 162 (1970)
5.32 M. Peshkin, I. Talmi, L. J. Tassie: Ann. Phys. (N.Y.) **12**, 426 (1961)
5.33 L. J. Tassie: Phys. Lett. **5**, 43 (1963)
5.34 A. Tonomura: Physics Today **43**, 22 (April 1990)
5.35 A. Tonomura: Intern. J. Mod. Phys. **3B**, 521 (1989)
5.36 A. Tonomura, H. Umezaki, T. Matsuda, N. Osakaba, J. Endo, Y. Sugita: In Proc. Intern. Symp. Foundations of Quantum Theory, ed. by S. Kamefuchi (Phys. Soc. Japan, 1984) p. 20
5.37 A. Tonomura, T. Matsuda, J. Endo, T. Arii, K. Mihama: Phys. Rev. Lett. **44**, 1430 (1980)
5.38 D. R. Tilley, J. Tilley: *Superfluidity and Superconductivity* (2nd ed.) (Hilger, Bristol 1986)
5.39 N. Byers, C. N. Yang: Phys. Rev. Lett. **7**, 46 (1961)
5.40 M. Peshkin: Phys. Reports **80**, No 6, 375 (1981)
5.41 F. London: Phys. Rev. **74**, 562 (1948)
5.42 J. M. Ziman: Principles of the Theory of Solids (2nd ed.) (Cambridge University Press, Cambridge 1979)
5.43 J. Bardeen, Phys. Rev. Lett. **7**, 162 (1961)
5.44 B. S. Deaver Jr., W. M. Fairbank: Phys. Rev. Lett. **7**, 43 (1961); R. Doll, M. Näbauer: Phys. Rev. Lett. **7**, 51 (1961)
5.45 W. L. Goodman, W. D. Willis, D. A. Vincent, B. S. Deaver Jr.: Phys. Rev. **B 4**, 1530 (1971)
5.46 G. N. Watson: *Bessel Functions* (2nd ed.) (Cambridge University Press, Cambridge 1958); A. Erdélyi, W. Magnus, F. Oberhettinger, F. G. Tricomi: *Higher Transcendental Functions*, Vol. II (McGraw-Hill, New York 1953)

Chapter 6

6.1 W. Heisenberg: *The Physical Principles of Quantum Theory* (transl. C. Eckart and F. C. Hoyt) (University of Chicago Press, Chicago 1930)
6.2 R. V. Jones: *Most Secret War* (Hamish Hamilton, London 1978)
6.3 M. Peshkin, A. Tonomura: *The Aharonov–Bohm Effect.* Lecture Notes in Physics **340** (Springer, Berlin, Heidelberg (1989)
6.4 B. D. Josephson: Phys. Lett. **1**, 251 (1962)
6.5 A. Baroni, G. Paternò: *Physics and Applications of the Josephson Effect* (Wiley, New York 1982)
6.6 J. Clarke: Nature (London) **372**, 501 (1994)
6.7 J. Clarke: Nature (London) **352**, 110 (1991)
6.8 G. Schön, A. D. Zaikin: Phys. Reports **198**, 237 (1990)
6.9 Y. Srivastava, A. Widom: Phys. Reports **148**, No 1, 1 (1987)

6.10 R. C. Jaklevic, J. Lambe, A. H. Silver, J. E. Mercereau: Phys. Rev. Lett. **12**, 159 (1964)
6.11 P. Carruthers, N. N. Nieto: Rev. Mod. Phys. **40**, 411 (1968)
6.12 J. R. Tucker, M. J. FeldmanL: Rev. Mod. Phys. **57**, 1055 (1985)
6.13 M. Tinkham: *Introduction to Superconductivity* (McGraw-Hill, New York 1975)
6.14 P. W. Anderson: In *Progress in Low Temperature Physics*, Vol. V, ed. by C. J. Gorter (North-Holland, Amsterdam 1967) pp. 1–43
6.15 P. W. Anderson: In *Lectures on Many Body Physics*, ed. by E. R. Caianiello, Vol. 2 (Academic Press, New York 1964) p. 113
6.16 R. C. Jaklevic, J. Lambe, J. E. Mercereau, A. H. Silver: Phys. Rev. **140A**, 1628 (1965)
6.17 D. Rogovin, M. Scully: Phys. Reports **25**, No 3, 175 (1976)
6.18 D. R. Tilley, J. Tilley: *Superfluidity and Superconductivity* (2nd ed.) (Hilger, Bristol 1986)
6.19 W. J. Ellon, M. Matters, U. Gelgenmüller, J. E. Mools: Nature **371**, 594 (1994)

Chapter 7

7.1 Y. Aharonov, D. Bohm: Phys. Rev. **115**, 485 (1959)
7.2 M. Peshkin, A. Tonomura: *The Aharonov–Bohm Effect*. Lecture Notes in Physics **340** (Springer, Berlin, Heidelberg (1989)
7.3 B. E. Allman, A. Cimmino, A. G. Klein, G. I. Opat, H. Kaiser, S. A. Werner: Phys. Rev. Lett. **68**, 2409 (1992)
7.4 G. Mateucci, G. Pozzi: Phys. Rev. Lett. **54**, 2469 (1985)
7.5 M. Peshkin: Phys. Rev. Lett. **69**, 2017 (1992)
7.6 J. Hamilton: *Theory of Elementary Particles* (Oxford University Press, Oxford 1959)
7.7 E. Merzbacher: *Quantum Mechanics* (Wiley, New York 1961) p. 273
7.8 W. H. K. Panofsky, M. Phillips: *Classical Electricity and Magnetism* (Addison-Wesley, Mass. 1955) (1st ed.) Sect. 17.2
7.9 Y. Aharonov, A. Casher: Phys. Rev. Lett. **53**, 319 (1984)
7.10 C. R. Hagen: Phys. Rev. Lett. **64**, 2347 (1990)
7.11 A. S. Goldhaber: Phys. Rev. Lett. **62**, 482 (1989)
7.12 M. Peshkin, H. J. Lipkin: Phys. Rev. Lett. **74**, 2847 (1995)
7.13 A. G. Klein: Physica **137B**, 230 (1986)
7.14 Yu. A. Sitenko: Sov. Journal of Nuclear Phys. (U.S.A.) 55(10), 1589 (1992)
7.15 B. E. Allman, A. Cimmino, A. G. Klein, G. I. Opat, H. Kaiser, S. A. Werner: Phys. Rev. **A 48**, 1799 (1993)
7.16 J. Anandan: Phys. Rev. Lett. **48**, 1660 (1982)
7.17 D. Greenberger, D. K. Atwood, J. Arthur, C. G. Shull, M. Schlenker: Phys. Rev. Lett. **47**, 751 (1981)
7.18 A. Zeilinger: *Fundamental Aspects of Quantum Theory*, ed. by V. Gorini, A. Frigerio, NATO ASI Ser. B, Vol. 144 (Plenum, New York 1986) p. 331
7.19 P. Pfeifer, A. G. Klein: Phys. Rev. Lett. **72**, 305 (1994)

Chapter 8

8.1 J. H. Hannay: J. Phys. A (Math. Gen.) **18**, 221 (1985)
8.2 M. V. Berry: J. Phys. A (Math. Gen.) **18**, 15 (1985)
8.3 G. Sagnac: Compt. Rend. **157**, 708, 1410 (1913)
8.4 G. Sagnac: J. Phys. Rad. (5th Series) **4**, 177 (1914)
8.5 L. D. Landau, E. M. Lifschitz: *The Classical Theory of Fields*, Vol. 2, 4th edn. (Pergamon, Oxford 1975) p. 254
8.6 E. J. Post: Rev. Mod. Phys. **39**, 475 (1967)
8.7 P. W. Forder: J. Phys. A (Math. Gen.) **17**, 1343 (1984)
8.8 M. V. Berry: Proc. R. Soc. Lond. **A 392**, 45 (1984)
8.9 T. Kato: J. Phys. Soc. Japan **5**, 435 (1950)
8.10 A. Messiah: *Quantum Mechanics*, Vol. 2 (North-Holland, Amsterdam 1962) p. 744
8.11 B. Simon: Phys. Rev. Lett. **51**, 2167 (1983)
8.12 V. I. Arnold: *Mathematical Methods of Classical Mechanics* (Springer, New York 1978)
8.13 E. Fues: *Handbuch der Physik*, Band V, Grundlagen der Mechanik (Springer, Berlin 1927) pp. 148–150
8.14 A. J. McConnell: *Applications of the Absolute Differential Calculus* (Blackie, London 1936)
8.15 R. W. Bowen, C-C Wang: *Introduction to Vectors and Tensors*, Vol. 2 (Plenum, New York 1976) p. 347
8.16 J. Roe: *Elementary Geometry* (Oxford Science Publications, Oxford 1993)
8.17 F. Harress: Thesis (Jena) (1911) unpublished
8.18 A. A. Michelson, H. G. Gale: Nature **115**, 566 (1925); Astrophys. J. **61**, 137 (1925)
8.19 S. A. Werner, J. L. Staudenmann, R. Colella: Phys. Rev. Lett. **43**, 1103 (1979)
8.20 P. Harzer: Astron. Nachr. **198**, 378 (1914)
8.21 M. Dresden, C. N. Yang: Phys. Rev. **D20**, 1846 (1979)
8.22 J. Anandan: Phys. Rev. **D24**, 338 (1981)
8.23 K-X Sun, M. M. Fejer, E. Gustafson, R. L. Byer: Phys. Rev. Lett. **76**, 3053 (1996)
8.24 T. L. Gustavson, P. Bouyer, M. A. Kasevich: Phys. Rev. Lett. **78**, 2046 (1997)
8.25 O. Avenel, E. Varoquaux: Phys. Rev. Lett. **60**, 416 (1988)
8.26 K. Schwab, N. Bruckner, R. E. Packard: Nature **386**, 585 (1997)
8.27 O. Avenel, E. Varoquaux: Czech Journal Physics (Suppl. S6) **48**, 3319 (1996)
8.28 C. W. F. Everitt: Nature **386**, 552 (1997)
8.29 M. V. Berry: Physics Today **43**, No. 12, 34 (1990)
8.30 A. Bohm, B. Kendrick: Int. J. Quantum Chem. **41**, 53 (1992)
8.31 A. Bohm, B. Kendrick, M. E. Lowe, L. J. Boye: J. Math. Phys. **33**, 977 (1992)
8.32 A. Shapere, F. Wilezek (eds.): *Geometric Phases in Physics* (World Scientific, Singapore 1989)
8.33 E. T. Copson: *Theory of Functions of a Complex Variable* (Oxford University Press, Oxford 1935) pp. 269–270
8.34 A. Erdélyi (Ed.): *Higher Transcendental Functions*, Vol. 2 (McGraw-Hill, New York 1953) pp. 188–192, 199, 204–205 (see also Vol. 3)
8.35 P. A. M. Dirac: *The Principles of Quantum Mechanics* (2nd edn.) (Oxford University Press, Oxford 1935)
8.36 M. S. Livingston: *High Energy Accelerators* (Interscience, New York 1954)
8.37 S. Appelt, G. Wäckerle, M. Mehring: Phys. Rev. Lett. **72**, 3921 (1994)
8.38 Y. Aharonov, D. Bohm: Phys. Rev. **115**, 485 1959
8.39 W. Ehrenberg, R. E. Siday: Proc. Phys. Soc. London **62B**, 8 (1949)

8.40 R. Y. Chiao, Y-S Wu: Phys. Rev. Lett. **57**, 933 (1986)
8.41 M. V. Berry: Nature **326**, 277 (1987)
8.42 S. M. Rytov: Dokl. Akad. Nauk. USSR **18**, 263 (1938)
8.43 M. Born, E. Wolf: *Principles of Optics* (Pergamon, London 1959)
8.44 F. Bortolotti: Rend. R. Acc. Naz. Linc. 6a, **4**, 552 (1926)
8.45 A. W. Snyder, J. P. Love: *Optical Wave Guide Theory* (Chapman and Hall, London 1983)
8.46 C. E. Weatherburn: *Differential Geometry in Three Dimension* (Cambridge University Press, Cambridge 1931)
8.47 M. P. Varnham, R. D. Birch, D. N. Payne, J. D. Love: Conf. Optical Fiber Commun., Atlanta, 68–71 (Optical Society of America 1986)
8.48 Proc. Int. Conf. On Fundamental Aspects of Quantum Theory, Columbia, S.C. (1989) (World Scientific, Singapore 1990)
8.49 J. N. Ross: Optical Quantum Electron. **16**, 455 (1984)
      M. P. Varnham, R. D. Birch, D. N. Payne: In Int. Conf. on Integrated Optics and Optical Fiber Commun. European Conf. on Opt. Commun., Geneva (1985) p. 135
8.50 A. Tomita, R. Y. Chiao: Phys. Rev. Lett. **57**, 937 (1986)
8.51 G. Delacrétaz, E. R. Grant, R. L. Whetten, L. Wöste, J. W. Zwanziger: Phys. Rev. Lett. **56**, 2598 (1986)
8.52 M. S. Smith, K. Budden: Proc. Royal Soc. A **350**, 27 (1975)
8.53 V. V. Vladimirskii: Dokl. Akad. Nauk. USSR **21**, 222 (1941); reprinted in B. Markovski, V. I. Vinitsky: Topological Phases in Quantum Theory (World Scientific, Singapore 1989)
8.54 I. and Z. Bialynicki-Birisla: Phys. Rev. **D 35**, 2383 (1987)
8.55 F. Wilczek, A. Zee: Phys. Rev. Lett. **52**, 2111 (1984)
8.56 A. Shapere, F. Wilczek (eds.): *Geometrical Phases in Physics* (World Scientific, Singapore 1989)
8.57 T. F. Jordan: J. Math. Phys. **28**, 1759 (1987)
8.58 H. Mathur: Phys. Rev. Lett. **67**, 3325 (1991)
8.59 J. E. Avron, L. Sadun, J. Segert, B. Simon: Phys. Rev. Lett. **61**, 1329 (1988)
8.60 C. Itzykson, J.-B. Zuber: *Quantum Field Theory* (McGraw-Hill, New York 1980)
8.61 O. Avenel, P. Hakonen, E. Varoquaux: Phys. Rev. Lett. **78**, 3602 (1997)
8.62 L. D. Faddeev, A. A. Slavnov: *Gauge Fields* (Benjamin/Cummings, London 1980)

Appendices

B.1 T. C. Northrop, E. Teller: Phys. Rev. **117**, 215 (1960)
B.2 Bo Lehnert: *Dynamics of Charged Particles* (North Holland, Amsterdam 1964) Chap. 4
C.1 P. M. Morse, H. Feshbach: *Methods of Theoretical Physics*, Vol. II (McGraw-Hill, New York 1953)
F.1 A. J. McConnell: *Applications of the Absolute Differential Calculus* (Blackie, London 1936)
F.2 R. W. Bowen, C-C Wang: *Introduction to Vectors and Tensors*, Vol. 2 (Plenum, New York 1976) p. 347
F.3 J. Roe: *Elementary Geometry* (Oxford Science Publications, Oxford 1993)
F.4 C. Itzykson, J-B. Zuber: *Quantum Field Theory* (McGraw-Hill, New York 1980)
G.1 Y. Aharonov, D. Bohm: Phys. Rev. **115**, 485 (1959)
G.2 S. Olariu, I. I. Popescu: Rev. Mod. Phys. **57**, 339 (1985)

G.3 M. Peshkin, I. Talmi, L. J. Tassie: Ann. Phys. (N.Y.) **12**, 426 (1961)

G.4 J. Hamilton: *Theory of Elementary Particles* (Oxford University Press, Oxford 1959)

G.5 M. Kretschmar: Z. Phys. **185**, 84 (1965)

G.6 E. Corinaldesi, F. Rafeli: Am. J. Phys. **46**, 1185 (1978)

G.7 M. V. Berry, R. G. Chambers, M. D. Large, C. Upstill, J. C. Walmsley: Eur. J. Phys. **1**, 154 (1980)

G.8 C. R. Hagen: Phys. Rev. Lett. **64**, 2347 (1990)

G.9 L. V. Hoang, L. X. Hai, L. I. Komarov, T. S. Romanova: J. Phys. A., Math. Gen. **25**, 6461 (1992)

H.1 L. D. Landau, E. M. Lifshitz: *Quantum Mechanics, Non-Relativistic Theory* (3rd ed.) (transl. J. B. Sykes and J. S. Bell) (Pergamon, Oxford 1977)

H.2 J. M. Ziman: Principles of the Theory of Solids (2nd ed.) (Cambridge University Press, Cambridge 1979)

H.3 L. D. Landau: Z. Phys. **64**, 629 (1930)

H.4 L. Page: Phys. Rev. **36**, 444 (1930)

H.5 E. T. Copson: *Theory of Functions of a Complex Variable* (Oxford University Press, Oxford 1935) pp. 269–270

H.6 A. Erdélyi (Ed.): *Higher Transcendental Functions*, Vol. 2 (McGraw-Hill, New York 1953) pp. 188–192, 199, 204–205 (see also Vol. 3)

Subject Index

Springer Tracts in Modern Physics